# Dynamics of Meteor Outbursts and Satellite Mitigation Strategies

# Dynamics of Meteor Outbursts and Satellite Mitigation Strategies

**Glenn E. Peterson**

The Aerospace Press • El Segundo, California

American Institute of Aeronautics and Astronautics, Inc. • Reston, Virginia

The Aerospace Press
2350 E. El Segundo Boulevard
El Segundo, California 90245-4691

American Institute of Aeronautics and Astronautics, Inc.
1801 Alexander Bell Drive
Reston, Virginia 20191-4344

Library of Congress Catalog Card Number 98-074939

ISBN 1-884989-06-3

*Let's burn down the observatory so this never happens again.*

Moe the bartender

# Contents

# Algorithms

# Tables

# Preface

Meteoroids are small particles produced by comets as they orbit about the Earth. When a comet approaches the Sun, it will heat up; small sandgrain-sized particles will then extrude off the surface leaving a cloud of debris in space. As time passes, the debris will spread into a ring around the Sun that roughly follows the comet's orbit. If the Earth goes through such a ring, a meteor shower is produced. Usually, the comets that create meteor showers as seen from the Earth have evaporated, leaving only the particle residue. The Leonid showers are unusual in that the comet that produces the particles, Tempel-Tuttle, is still active. It orbits the Sun once every 33 years, producing intense periods of activity with the same period, give or take a year. Because Tempel-Tuttle crossed the orbit of the Earth in the spring of 1998, the Leonid activity is potentially several orders of magnitude greater than usual in 1998–2000. The 1997 shower produced one order of magnitude greater intensity than usual; as this book was being written, the 1998 Leonids produced intensities of about 20 times greater than normal.

These meteoroid particles have been a threat to spacecraft in the past. In 1991, the Japanese ISAS spacecraft Solar A experienced a loss of its optical telescope as a result of an impact from the Perseid shower. In 1993, the same shower caused the European Space Agency's Olympus I telecommunications satellite to spin out of control. While the vehicle was not destroyed directly, so much fuel was used in regaining attitude control that the mission was effectively ended. As a consequence, meteoroid particles have been a historical source of severe satellite mishaps, and with the greater intensities expected for the Leonids in the next few years, they will continue to be a danger.

Given this threat, it is of value to the space community to have available to them a document that describes what can be done to protect their spacecraft. This book summarizes the meteoroid threat, with special emphasis on, but not limited to, the upcoming Leonids. It examines the level of shadowing produced by the Earth that spacecraft can expect to acquire. How the Earth's gravity focuses the meteors and how the changed trajectories alter our perceptions of the problem are also studied. Most significantly, several mitigation options are analyzed:

- Deriving optimal attitude control to minimize the cross-sectional area of the vehicle, solar panels, or the sensors
- Altering attitude control to lessen the produced spall and charge
- Altering the orbit to reduce the impact damage
- Altering the orbit to reduce the ambient flux
- Changing the plane of the orbit to take advantage of the Earth's shadow

Fuel consumption, as well as the dynamic analysis of the mitigation options, is conducted. The study is general in nature and therefore applicable to any meteor threat, present or future. The results are presented in a series of algorithms that allow the reader to easily implement the procedures. The specific applications and numerical results are tailored for the Leonids.

# Acknowledgments

No researcher can succeed without the help and assistance of many others. The writing of a book takes much time, effort, and patience not only of the writer but of others as well. My many thanks go to everyone who assisted in this undertaking whether they were aware of their help or not. However, there are several people that I wish to thank individually for their contribution to this endeavor:

First and foremost, I want to deeply thank David Lynch of The Aerospace Corporation for providing the guidance for this research and for having the confidence in my abilities to let me run to the best of my abilities with this particular aspect of the Leonid meteor problem.

Tom Lang, Bob Williamson, Dwight Abbott, and the personnel of the Astrodynamics Department of The Aerospace Corporation for their support and partnership in the many interesting investigations that we have performed together.

I wish to especially thank my editor, Donna Born of The Aerospace Press, for going above and beyond the call of duty in the many rewritings that a sometimes finicky author required.

Hans Karrenberg for assigning me to David Lynch for the Leonid project for the simple reason that Dave and I are both from the University of Texas, Austin.

Ed Tagliaferri and the United States Air Force for their collaboration on the research. Without their foresight and warning in anticipating the Leonid problem, the satellite community would not have been as prepared for this potential disaster as it actually was.

The Aerospace Corporation; Bill Ailor, principal director of the Center for Orbital and Reentry Debris Studies; Dave Evans, executive director of The Aerospace Institute; and Velma Senaga of the Publications Department for providing the necessary support to see this project through.

I am indebted to the participants of the Leonid Meteor Storm and Satellite Threat Conferences for their many interesting and enlightening discussions on topics relating to the meteoric phenomena.

My family for their continuing support for my career and their help in all aspects of my life.

The American Institute of Aeronautics and Astronautics and The Aerospace Press for copublishing this book and allowing it to become available to a wide audience.

Finally, my gratitude goes to everyone not listed above who has contributed to this research. Memories will fade over time and it is hard to keep track of all those who have helped in such an effort as this book has taken. My thanks go to all of them and my apologies as well for missing any who warranted individual mention.

# 1 Assessing the Threat

Meteoroid impacts have represented a very real threat to satellites in orbit about the Earth in the past and certainly will into the future. This chapter quantifies the dynamics of this threat. The particle's orbit about the Sun is computed, and an accurate location of the radiant point (apparent direction from which the particles come when viewed from the Earth) associated with the computed orbit is estimated. This is followed by an analysis of the expected particle fluxes. The analysis to be performed will be general in nature in order to be applicable to any future meteor event, but the examples given throughout this document will emphasize the 1998–1999–2000 Leonids since that scenario poses the greatest threat at the present time.

## 1.1 Death from Above

Several spacecraft have been known to suffer from meteorite hits, resulting in either catastrophic loss of the vehicle or significant degradation of mission performance. The Perseid meteor showers of 1991 and 1993 caused effective loss of the ISAS Solar A and ESA Olympus I satellites, respectively. The vehicles were not destroyed through a catastrophic impact; rather, their mission effectiveness was degraded in both cases to such an extent that the mission was essentially over. Solar A experienced a puncture of its telescopic shade, which ruined the main experiment. Olympus I experienced erroneous attitude control, causing virtually all the fuel to be expended by the time control was reestablished (Caswell, 1998). NASA's International Sun-Earth Explorer (ISEE) also experienced a degradation of mission goals because of a sporadic meteor hit not associated with any known shower. While meteor strikes are rare, their impact can be disastrous on the mission.

Meteor showers that have the potential to damage spacecraft occur when the Earth intercepts a stream in space consisting of small meteoroid particles. The edges of the debris tube are diffuse and ill-defined; just when the Earth enters and leaves the stream is thus somewhat arbitrary. However, the general concept of the Earth traveling through a tube of small cometary particles holds (Fig. 1.1). The closer the Earth gets to the center of the stream, the more intense the meteor activity is.

These meteoroids are created by a parent body—either a comet or less likely an asteroid—in a continuous, although nonconstant, process as the parent moves about the Sun. As a comet comes closest to the Sun at periapse, the increase in solar heating causes more and more ice to vaporize and extrude from the surface of the body. At periapse, particle production thus reaches a maximum. As the particles leave the surface, they begin to distribute themselves out along the path of the comet, creating a ring of meteoroids somewhat akin to the rings of Saturn. Eventually, they smooth themselves out so that the particle density is similar throughout the ring. However, for meteor streams whose parent has not yet completely evaporated or broken up, the particle density is, of course, largest close to the parent when that body in turn is closest to the Sun. If the Earth happens to pass close by the parent when the parent crosses the plane of the ecliptic, the resulting meteor shower has the potential of being more intense than usual. If this

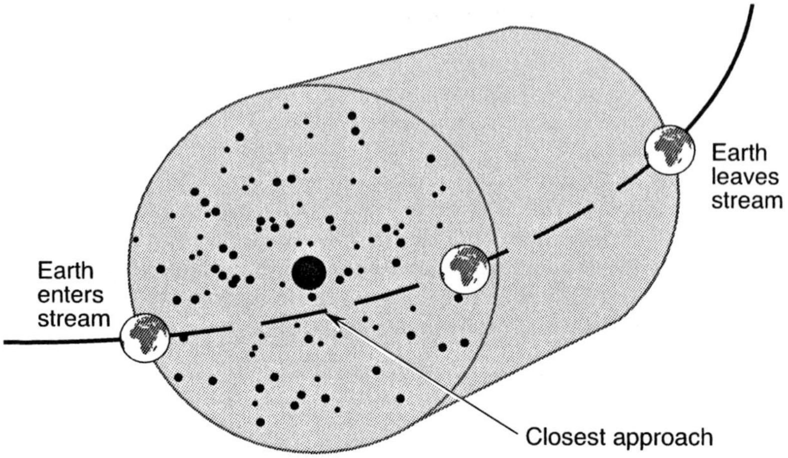


**Fig. 1.1. Earth traveling through a meteor stream tube.**

greater concentration takes a long time to disperse (i.e., multiple cometary periods), then each passage of the comet close to perihelion will create a "streamlet" within the larger stream that is still fairly close to the parent comet. These greater concentrations, either due to streamlets or fresh material, will be referred to as a near-comet type of meteor outburst. Far-comet types of outbursts can occur as well, wherein the generating comet is far from the Earth, but the resulting stream is perturbed such that either the Earth travels closer to the center of the stream or a resonance effect focuses the particles into a region of greater density. A greater-than-normal level of activity thus results. Regardless of whether the outburst is of a near-comet or far-comet type, normally tame meteor showers can turn into monstrous affairs for satellite operators (Fig. 1.2).

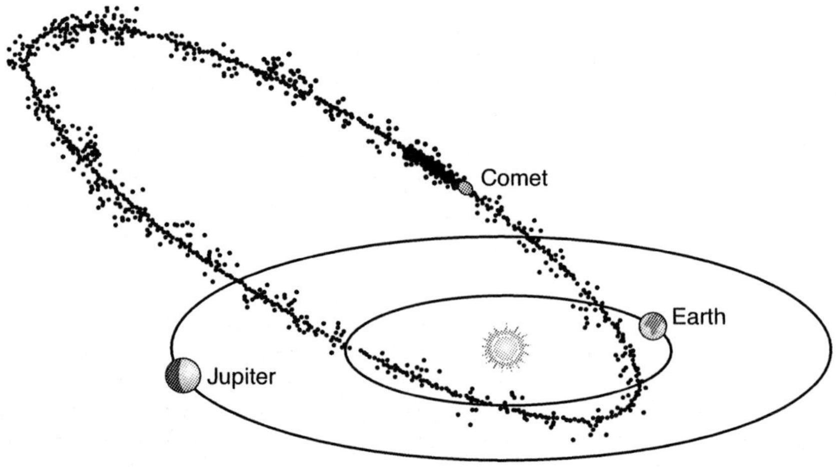


**Fig. 1.2. Particle density as a function of comet location.**

This situation has occurred in the past for various meteor showers and is happening again now for the Leonids. The Leonids' parent, comet P55/Tempel-Tuttle, cycles through its orbit once every 33 years. It most recently passed the Earth's orbit plane in early 1998; as a consequence, the Leonid showers of November 1998–2000 are expected to peak at an intensity that may be several orders of magnitude greater than usual. A similar situation occurred the previous time Tempel-Tuttle came around, in 1966. The resulting storm was anywhere from 1000 to 10,000 times greater in intensity than normal. Because there were so few satellites in orbit at that time, the storm did not cause any noticeable effects on the spacecraft. However, with so many satellites aloft at the current time and even though the 1998–2000 events are not expected to be as intense as they were in 1966, the anticipated increase in activity will pose a significant risk to orbiting vehicles.

## 1.2 Orbit Dynamics

Because the nature of meteor outbursts is governed by how close the comet is to the Earth at meteor stream encounter and how close the Earth is to the center of the stream, it is imperative to understand the basic dynamics of the system. This section examines the parent comet's orbit, the sublimated particles' orbit, and the time it takes for the newly created particles to disperse throughout the stream into an even distribution.

This book assumes a basic knowledge of orbital mechanics and dynamics, but to refamiliarize the reader, an orbit is defined by these six parameters:

- Semimajor axis $a$
- Eccentricity $e$
- Inclination $i$
- Longitude of ascending node $\Omega$ (also called the right ascension of ascending node when referring to Earth-centered spacecraft)
- Argument of periapse $\omega$
- True anomaly $f$

This constitutes the classical Keplerian set. The semimajor axis and eccentricity define the size and shape of the orbital ellipse; the inclination, node, and argument of periapse set the orientation of the orbit in space (Fig. 1.3). The inclination is the angle the orbit makes with either the ecliptic plane (Sun-centered orbits) or the equatorial plane (Earth-centered). The ascending node location is the angle between the vector pointing from the central body to the first point of Ares ($x$-axis) and the location where the orbit crosses the ecliptic or equatorial plane moving from south to north ($z$-axis is perpendicular to this plane). The argument of periapse is the angle from the ascending node to the periapse location. Finally, the in-plane location of the satellite or comet is given by the true anomaly. For near-circular orbits, the eccentricity becomes small, resulting in the periapse location becoming ill-defined. In this case, the argument of periapse and the true anomaly are replaced by the more well-defined argument of latitude $u$ $(= \omega + f)$. Another way of indicating the in-plane location is through a fictitious angle, the mean anomaly $M$, which is linearly related to the mean motion (i.e., the semimajor axis) of the central body through Kepler's Equation. These orbital elements will be discussed repeatedly in this text. For

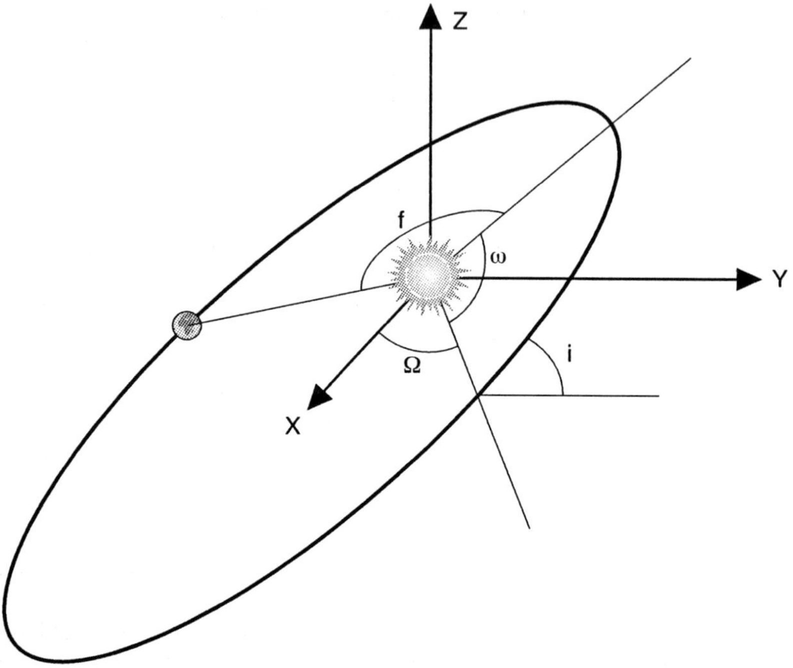


**Fig. 1.3. Basic orbit element angles.**

a more in-depth treatment of these definitions and the vast field of orbital mechanics, the interested reader is referred to additional sources in astrodynamics (Bate, Mueller, and White, 1971, and Vallado, 1997).

### 1.2.1 Parent Orbit

Because meteoroids are created by a parent comet, it is often useful as a first approximation to model the movement of the meteoroids as following the orbit of the parent body. But in modeling the motion of an arbitrary comet, the Sun is not the only gravitating body that needs to be considered. Other large bodies in the solar system can significantly influence periodic comets as well. Jupiter and Saturn, most notably, can create resonant types of effects that alter the behavior of either the comet or the generated dust stream. One theory holds that the rare storms associated with certain meteor streams that occur when their parent comets are far from perihelion are caused by a resonance within the dust cloud dictated by Jovian and Saturnian perturbations, which are far-comet types of outbursts (Jenniskens, 1995). The planetary-disturbing function must therefore include all significant bodies (i.e., Jupiter and Saturn, certainly) and possibly ones whose importance may not be immediately apparent. While the Earth, for example, is not an important gravitating body for virtually any comet, if a close approach to the Earth occurs, it can be quite significant in the resulting trajectory of that comet. Care must then be taken when excluding the smaller bodies. Note also that problems can develop during very close approaches to various third bodies. Some comets, such as Shoemaker-Levy 9, are more appropriately treated as in the influence of Jupiter as the dominant central body for at least a portion of their orbits. Such switching between central bodies (in this case, Jupiter and the Sun) creates special problems in the numerical integration, of which analysts must be aware (Vallado, 1997).

Simple relativistic effects also are typically included in the reduction of observational data. Because the best available ephemerides for the planets and the Sun include general relativity, small errors can result in the reduction of the comet's orbit if the same effect is not included in the comet's force modeling when the reduction occurs. Owing to the small sizes of the relativistic perturbations, only the Sun's additional curvature on the fabric of space time is necessary (Shahid-Saless and Yeomans, 1994).

In addition to the third-body perturbations and the central-body relativistic effect, comet orbits are also best simulated when a nonconservative force based upon Whipple's ejection model is used (Yeomans and Chodas, 1989). Whipple's model assumes that particle ejection occurs in a rocket-like manner as a result of solar heating of a rotating icy ball. If the nucleus of the comet were not spinning, the heating would be symmetrical about the radial axis (comet-Sun vector). But with a rotating snowball, a thermal lag angle develops between the point of maximum ejection and the subsolar point. Generally speaking, the ejection force will thus have transverse and normal components and not just the radial aspect. However, the normal component is small and periodic in nature and thus tends not to influence the long-term behavior of the given comet. Taken together, the equations of motion for a given comet are best modeled by

$$\frac{d^2\bar{r}}{dt^2} = -\mu_{\text{sun}}\frac{\bar{r}}{r^3} + \sum_{j=1}^{N}\mu_j\left[\frac{\bar{r}_j-\bar{r}}{\|\bar{r}_j-\bar{r}\|^3} - \frac{\bar{r}_j}{\|\bar{r}_j\|^3}\right] + g(r')[A_1\hat{R} + A_2\hat{T}] \tag{1.1}$$
$$+\frac{\mu_{\text{sun}}}{r^3}\left\{\left[\frac{2(\gamma+\beta)}{c^2}\frac{\mu_{\text{sun}}}{r} - \frac{\gamma}{c^2}v^2\right]\bar{r} + \left[\frac{2(\gamma+1)}{c^2}(\bar{r}\cdot\bar{v})\right]\bar{v}\right\}$$

where $\mu_{\text{sun}}$ is the Sun's gravitational parameter, $\gamma$ and $\beta$ are relativistic parameters describing the curvature of space time (both are equal to unity in general relativity), $c$ is the speed of light, $N$ is summed over all the required planetary bodies (note again that very close approaches may require a different third-body formulation), $r$ and $v$ are the comet's heliocentric position and velocity vectors, and the index $j$ is summed over all the necessary planetary bodies. The third term of Eq. (1.1) depicts Whipple's icy snowball model, where $R$ and $T$ are unit vectors in the radial and transverse directions in the plane of the comet's orbit, and $g$ is a function describing the aqueous physical composition of the comet:

$$g(r') = \alpha\left(\frac{r}{r_0}\right)^{-m}\left[1+\left(\frac{r}{r_0}\right)^n\right]^{-k} \tag{1.2}$$

and

$$r' = r(t) = r(t - DT) \tag{1.3}$$

where $DT$ denotes a comet-specific asymmetry in the cometary ejection about perihelion passage; $m$, $n$, $k$, $r_0$, and $\alpha$ are parameters dictated by the comet's physical nature (i.e., icy snowball) and therefore are assumed invariant between individual comets. $A_1$ and $A_2$ are comet-specific values determined from observations for the radial and transverse components of the acceleration (the normal component is again typically insignificant). Because they are empirically determined from measurements, $A_1$ and $A_2$ tend to absorb some of the other nonconservative forces, such as solar radiation pressure, that would otherwise act in the radial and transverse directions. This is the traditional approach to generating cometary orbits. The reader should be aware that when tabulated values of $A_1$ and $A_2$ appear, the above force model (Eq. [1.1]) has been used. Adding in solar radiation pressure or other forces on the comet as well as the empirical ejection forces could therefore generate a slightly erroneous orbit.

Table 1.1 shows values typical for Whipple's ejection model. The first two columns are parameters dictated by the assumption of the icy-snowball model and are assumed applicable for all comets. The third column contains parameters specific to each comet ($A_1$ and $A_2$ are in AU/day$^2$). The values shown are for comet Tempel-Tuttle, the parent body of the Leonids. The orbit elements for Tempel-Tuttle are additionally shown in Table 1.2. The orbit elements of some other significant comets (at least as far as meteor showers are concerned) are given in Appendix A. The orbit elements presented here were obtained from the Solar System Exploration Division of the NASA Jet Propulsion Laboratory.

For near-comet types of meteor outbursts, the intensity of encounter is driven by two factors—how close typically the comet's (or streamlet's) orbit is to the Earth's orbit at closest approach (a positive value indicates that the Earth passes inside the comet's orbit) and how much time has elapsed between the Earth's passage and the comet's passage of that point (a positive value indicates that the Earth is behind the comet). Although the stream particles will have a distinct distribution—it has been shown through modeling (Brown and Jones, 1993) that it may resemble a twisted braid-like structure—the exact distribution is not

**Table 1.1. Comet Tempel-Tuttle Model Parameters**

| | | | | | |
|---|---|---|---|---|---|
| m | 2.15 | $r_0$ | 2.808 | $A_1$ | 5.4 (±2.7) e-10 |
| n | 5.093 | $\alpha$ | 0.111262 | $A_2$ | 9.129 (±0.015) e-11 |
| k | 4.6142 | | | DT | 0.0 |

**Table 1.2. Comet Tempel-Tuttle Keplerian Elements for Epoch 1998 Mar 8.0**

| | |
|---|---|
| Perihelion distance (AU) | 0.9765849 |
| Eccentricity | 0.9055036 |
| Inclination (deg) | 162.48614 |
| Longitude of ascending node (deg) | 235.25883 |
| Argument of perihelion (deg) | 172.49731 |
| Time of perihelion passage | 1998 Feb 28.09666 |

determinable before the encounter occurs. Therefore the only way to predict whether an encounter may be intense is through studying the dynamics of the encounter.

By comparing the ephemeris found from Eq. (1.1) with that of the Earth's orbit, the point of closest approach can be calculated for any encounter. Table 1.3 shows the data for the Leonid storms that were computed using the above model. The zenith hourly rate (ZHR) is a measure of the observational intensity of a meteor stream's activity (discussed in further detail in Sec. 1.4). Historical values of the ZHR are taken from recent sources (Jenniskens, 1994 and Yeomans *et al.*, 1996). It should be noted that the values given here are for the peak returns for that particular return of Tempel-Tuttle. They do not necessarily correspond to the years where the Earth passes closest behind the comet. For example, the Earth passed just 129 days behind the comet in 1899, but that storm had a ZHR estimated in the 1100 range only. Two years later, the 1901 storm was quite a bit more active than that, with a peak ZHR of approximately 7000. In 1903, 4 years after the comet's passage, the Leonids again had a ZHR over 1000. It is important to note that just because the Earth is close to the comet temporally does not mean that a storm will take place, nor does it mean this encounter will be the most active for that particular comet passage. A more telling measure of activity than closeness in time is the distance of closest approach. But even that comes with a caveat. The 1932–33 encounter was spatially closer to the center of the stream than the 1899–1901–1903 events, but was disappointing in its behavior. These discrepancies are representative of the complexity and uncertainty of evaluating meteoric events.

### 1.2.2 Particle Ejection

Once the basic orbit of the meteor stream is established based upon the behavior of the parent comet, the ejection of the future meteoric particles must commence. Although near-comet outbursts can be modeled based primarily upon the orbit of the parent, general meteor stream behavior might not follow the parent closely. Streamlets generated during past perihelion passages will follow the parent orbit closely but will deviate slightly over time. Far-comet outbursts such as the Lyrids of 1992 can behave even more differently, and some streams are not

**Table 1.3. Tempel-Tuttle Earth Encounter Data from Past Leonid Storms**

| Year | Distance (AU) | Time (days) | ZHR |
|---|---|---|---|
| 1799 | –0.0034 | –116 | >5000 |
| 1833 | –0.0012 | +307 | >5000 |
| 1866 | –0.0050 | +298 | ~17,000 |
| 1901 | –0.0168 | +859 | ~7,000 |
| 1933 | –0.0118 | +126 | <1000 |
| 1966 | –0.0035 | +561 | 15,000–150,000 |
| 1998–2000 | –0.0085 | +257 | ~10,000 (expected) |

associated with any known parent at the current time (Quadrantids). The particle ejection behavior is thus significant because that provides a fundamental, random addition to the initial velocities of the particles when computing the subsequent evolution of the meteor stream.

Consider a particle that has broken away from the parent body. Such a particle is far enough away from the cometary nucleus that the specific sublimation mechanism that created the particle as being separate from the comet body can be ignored. In this case the future meteoroid will have momentum imparted to it from the more highly energetic gas that is sublimating off the comet due to the solar heating over the hemisphere that is facing the Sun. When this sublimating force is larger than the local gravity acting to pull the particle back to the surface, the particle will escape and become part of the meteor stream (Fig. 1.4).

Assuming the two forces act purely radially, the velocity computed at infinity (i.e., the delta to be added to the initial velocities) (Whipple, 1951) is:

$$v_{\text{eject}}^2 = \frac{13 v_{\text{gas}} F_{\text{sun}} R_c}{12 n_{\text{eff}} r^2 s_p \rho_p H} - 2\frac{\mu_c}{R_c} \tag{1.4}$$

where $F_{\text{sun}}$ is the solar constant (315 cal/sec at 1 AU), $\mu_c$ (km$^3$/sec$^2$) and $R_c$ (km) are the gravitational parameter and radius of the parent comet, $r$ is the solar distance in AU, $s_p$ (cm) and $\rho_p$ (g/cm$^3$) are the radius and density of the particle, $H$ is the mean heat of sublimation (~450 cal/g), $1/n_{\text{eff}}$ represents the fraction of the solar heating that goes into sublimation ($1/n_{\text{eff}}$ can be assumed as unity for most comets; however, it can be much smaller: comet Encke's fraction is assumed to be ~0.1), and $v_{\text{gas}}$ is the average speed of the gas carrying the particle away (Whipple, 1950):

$$v_{\text{gas}} = \frac{0.56 \times 10^{-5}\text{cm/sec}}{r^{1/4}} \tag{1.5}$$

where the constant in the numerator is found as a function of the temperature and the gas constituents typical of comets. The magnitude of the velocity to be

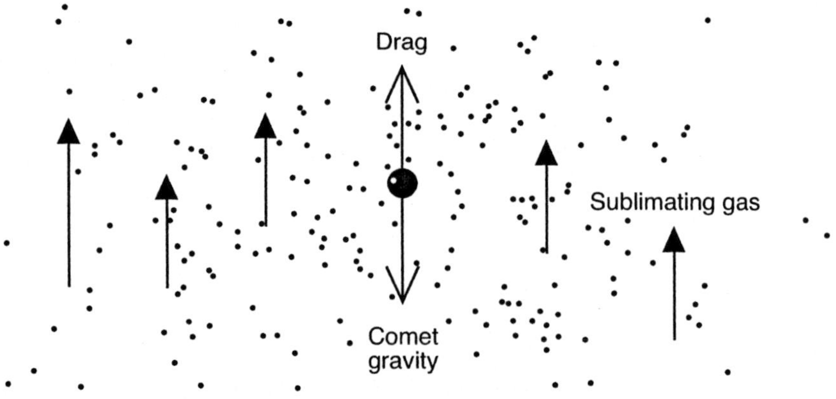


**Fig. 1.4. Forces acting on a particle in the near-comet environment.**

added in random directions to the particles' initial conditions for a typical comet is:

$$v_{\text{eject}} = \left(\frac{4R_c}{n_{\text{eff}} s_p r^{9/4}} - 0.052R_c^2\right)^{1/2} \times 0.00328 \text{ km/sec} \tag{1.6}$$

Once again, in this expression $R_c$ is in kilometers, $s_p$ is in centimeters, and $r$ is in AU. Note that there is a factor of 4 difference in the first term of this expression and one noted elsewhere (Whipple, 1951). The earlier model assumed that the meteor grains embedded in the icy comet were rocky in nature, resulting in a particle density of 4 g/cm$^3$. More recent studies (McDonnell *et al.*, 1991) show that the comet particles vary in composition from ice fragments to fluffy silicates. The particle densities thus range from 1 to 2.5 g/cm$^3$. Because the lower density gives a higher ejection velocity (and hence greater perturbation), the 1 g/cm$^3$ value is assumed in Eq. (1.6).

This perturbation on the initial velocity acts radially away from the comet nucleus. However, comets will be spinning as they move through their orbit. There will therefore be an additional transverse perturbation to the initial velocities of the particle cloud. This creates the braid-like structure in the resulting stream that was mentioned earlier. But current determinations of actual comet spin rates and directions are highly uncertain and are currently best left to the realm of theoretical modeling.

### 1.2.3 Ring Creation

Once the particles have escaped from the comet's influence, the small deviations in the velocities from that of the parent body will cause the particles to continue to move away from the basic orbit. The stream will disperse but remain fairly centered on the parent comet's orbit as long as the parent comet is in a fairly short period (several Jovian cycles or less). For these parents, the orbits of the child particles do not have time to be significantly dispersed or altered before a fresh batch of children is birthed by the parent. For long-period comets, substantial deviations can occur, resulting in long-period streams that may be quite different in their behavior than the parent comet.

Other than these small changes in the initial conditions, the most significant forces that cause the subsequent evolution of the particles to deviate from the parent trajectory are solar radiation pressure and Poynting-Robertson drag. Resonances can also impact ring evolution, but they are beyond the scope of this analysis. Recall that for the comet, the dominant nonconservative force is the rocket-like ejection force governed by Eq. (1.2). The other smaller nonconservative forces that must act on the comet were absorbed into the estimation of the empirically determined parameters $A_1$ and $A_2$. For the particles themselves, no such estimation can be made because the individual particles cannot be tracked. Therefore, in Eq. (1.1) the ejection force must be replaced by explicit expressions for the solar radiation pressure, the Poynting-Robertson drag, and the solar wind drag, as in Eq. (1.7).

$$\bar{F}_{\text{noncons}} = \beta\frac{\mu_{\text{sun}}}{r^2}\left[\left(1 - \frac{v_R}{c}\right)\frac{\bar{r}}{r} - \left(\frac{1 + sw}{c}\right)\bar{v}\right] \tag{1.7}$$

where $sw$ is the ratio of the solar wind drag to Poynting-Robertson drag (~0.3), $c$ is the speed of light, $r$ and $v$ are the position and velocity of the particle, $v_R$ is the velocity component in the radial direction, $\mu_{sun}$ is the gravitational parameter of the Sun, and β is the ratio of the solar radiation force to the gravitational force, valid for particles larger than 1 mm in radius (Dohnanyi, 1978):

$$\beta = 5.73 \times 10^{-5} c_R (\rho_p s_p)^{-1} \tag{1.8}$$

where $c_R$ is the coefficient of reflectivity for the particle and will vary from 0 (translucent material) to 1 (perfect absorber) and up to 2 (perfect reflector). It should be noted that very small particles do not behave according to this equation, but they do not constitute a threat to spacecraft. The Poynting-Robertson drag is small in comparison to the other forces but significant in that this force, for short-period comets, tends to force the smaller, nonspacecraft-threatening particles to spiral into the Sun and so acts to "clean up" the stream of very small dust.

There are other nonconservative forces that can act on particles in addition to solar radiation pressure, Poynting-Robertson drag, and solar wind drag. Yarkovsky forces can be as large as the Poynting-Robertson effect under certain circumstances (Burns *et al.*, 1979) but only for particles that are too small to be a threat to spacecraft. This force acts as a thrust on the particle along its spin axis because of the differential heating and dissipation that occurs when a rotating body is subjected to thermal radiation. Differential Doppler and Lorentz forces can also influence the behavior of particles but only in a very limited fashion. These smaller forces are typically ignored in stream particle studies. However, the Lorentz force is important for particles still in the coma of the parent (Horanyi and Mendis, 1991).

The dispersive solar radiation force and the small changes in the initial conditions experienced by the particles are the main drivers in determining how fast the larger, more dangerous particles spread out along the stream and thus complete the closure of the meteors into a continuous ring. Because the meteoroids are pushed both in front of and behind the comet by the ejection, the number of orbital revolutions required for the perturbed particles to go through one-half of the parent body's period defines the problem (McIntosh, 1973) as:

$$N_{rev} = \frac{1}{2}\frac{TP}{dTP} \tag{1.9}$$

where $TP$ is the period of the comet and $dTP$ the deviation from the comet's period experienced by the perturbed particle ($a$ is the semimajor axis of the orbit). The period and its derivative are given by:

$$TP = 2\pi\sqrt{\frac{a^3}{\mu_{sun}}} \qquad dTP = 3\pi\sqrt{\frac{a}{\mu_{sun}}}da \tag{1.10}$$

Assuming that the particle ejection takes place impulsively at perihelion, the differential changes for the two main perturbing forces (ejection and solar radiation pressure) that act to change the semimajor axis are:

$$da_{\text{eject}} = \frac{2a^2 v_c}{\mu_{\text{sun}}} v_{\text{eject}} \qquad da_{\text{srad}} = \frac{a^2 v_c^2}{\mu_{\text{sun}}} \beta \tag{1.11}$$

where $v_c$ is the speed of the comet at perihelion and $e$ is the eccentricity, producing

$$v_c = \sqrt{\frac{\mu_{\text{sun}}}{a} \frac{1+e}{1-e}} \tag{1.12}$$

Substituting, the number of revolutions it takes to close the orbit and create a continuous meteor stream is then (where in this expression the semimajor axis is written in AUs, the velocities are in km/sec, and $\mu_{\text{sun}}$ is in km$^3$/sec$^2$):

$$N_{\text{rev}} = \frac{1}{6a} \sqrt{\frac{\mu_{\text{sun}}}{149.5 \times 10^6}} \left[\frac{2}{a(1-e)} - \frac{1}{a}\right]^{-1/2} \left[v_{\text{eject}} \pm \frac{1}{2}\beta v_c\right]^{-1} \tag{1.13}$$

Note that the ejection velocity can act in either direction; however, in order to determine the minimum amount of time to complete closure, the maximum value of the combined velocity term should be taken.

Because both the ejection velocity and parameter $\beta$ are functions of the relatively uncertain comet physical properties as well as being highly dependent upon the particle size, the number of revolutions is not an easy matter to express. Figure 1.5 exhibits this property and shows the number of revolutions for several prominent meteor showers as a function of the particle size, assuming the cometary physical properties are constant ($\rho_p$ = 1 g/cm$^3$, $c_R$ = 1, $n_{\text{eff}}$ = 1). The time it takes to close a meteor stream into one continuous loop thus varies for

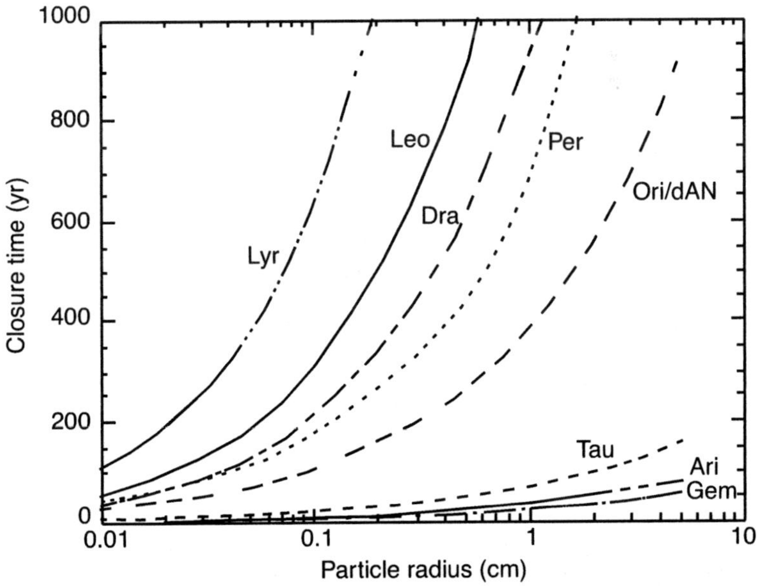


**Fig. 1.5. Complete stream formation time.**

different particle sizes. In theory, it may therefore be possible to compute the location of an unknown parent comet not only from the changes in the numbers of meteors that an Earth observer might see, but also from the year-to-year variations in the magnitudes of the observed meteor streaks.

While particles smaller than 0.01 cm may cause damage to spacecraft, that value was chosen as a limit in the figure not only for presentation purposes but also because for the fastest-moving meteors (i.e., Leonids at 70 km/sec), a particle of mass approximately $1 \times 10^{-5}$ g will create a magnitude 6.5 streak; slower particles require an even larger radius (see Sec. 1.4). With a particle density of 1 g/cm$^3$, a Leonid +6.5 magnitude–sized particle translates into a radius of 0.013 cm. Therefore, only under rare circumstances will particles smaller than 0.01 cm create noticeably visible meteor streaks in the night sky.

## 1.3 Radiant Point

With the stream dynamics now defined, the encounter at Earth crossing must be examined. One of the important characteristics for an Earth encounter is the radiant point of the resulting meteor event. The radiant point is the apparent location in the sky that the meteors come from when viewed by an observer on the Earth. Assuming the meteoroids generated by a comet closely follow the comet's orbit, the radiant point is simply computed from the relative velocity components between the meteor particles and the Earth at the point of closest approach between the Earth's orbit and the comet's orbit (Fig. 1.6). Note that the relative velocity must be taken from the Earth to the comet.

$$\bar{v}_{rel} = \bar{v}_{Earth} - \bar{v}_{comet} \tag{1.14}$$

The relative velocity at that point is then rotated into the geocentric equatorial plane with the right ascension and declination subsequently withdrawn from the geocentric relative velocity components as (Eq. [1.15]):

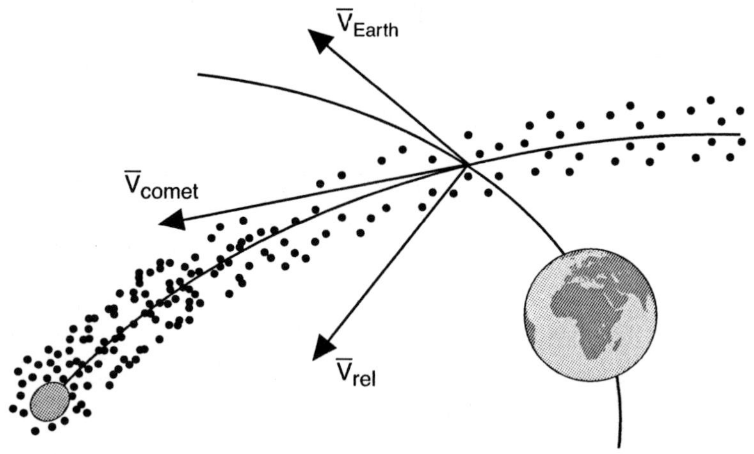


**Fig. 1.6. Relative velocity between the Earth and the comet at closest approach.**

$$\alpha_L = \tan^{-1}\left(\frac{v_{\text{rel }y}}{v_{\text{rel }x}}\right) \qquad \delta_L = \sin^{-1}\left(\frac{v_{\text{rel }z}}{\|v_{\text{rel}}\|}\right) \tag{1.15}$$

where the right ascension must be tested against the $x$- and $y$-components of the relative velocity to make sure the quadrant is correct and where the magnitude of the relative velocity is simply the approach velocity of the meteors with respect to the Earth (hereafter referred to as $V_\infty$). For the Leonids, the relative velocity vector as computed from the comet orbit in Sec. 1.2 at Earth crossing in km/sec is shown in Eq. (1.16).

$$\bar{v}_{\text{rel}} = [-58.8, 29.4, 26.3] \tag{1.16}$$

The results of the radiant point computation using the generated ephemeris for Tempel-Tuttle are given in Table 1.4, along with the same computation for several different currently available ephemerides. The newer ephemerides all give very close results, which instill confidence in the methodology employed here. A formal error study shows that the errors on the radiant points can be conservatively estimated at less than 0.1 deg.

The best geographical viewing locations are tied to the time of closest approach (from Sec. 1.2) and the radiant vector. Table 1.5 gives these values for the Leonids over the 1998–2000 time span.

**Table 1.4. Leonid Radiant Points from Various Sources**

| | Right Ascension (deg) | Declination (deg) |
|---|---|---|
| Peterson, 1998 | 153.45 | 21.87 |
| Yeomans, 1997 | 153.45 | 21.88 |
| Yeomans, 1996 | 153.34 | 21.76 |
| Muraoka, 1997 | 153.43 | 21.88 |
| MPC catalog (Marsden), 1997 | 153.42 | 21.88 |
| Yeomans, 1981 | 152.66 | 22.00 |
| LTEP (Carusi), 1985 | 152.69 | 22.15 |
| Cook, 1973 | 152.30 | 22.20 |

**Table 1.5. Leonid Visibility with Best Satellite Longitude**

| Date (± 2 h) | Best Visibility (deg E) |
|---|---|
| 1998 Nov 17: 20.64$^{\text{h}}$ UT | 130 ± 30 |
| 1999 Nov 18: 03.36$^{\text{h}}$ UT | 36 ± 30 |
| 2000 Nov 17: 10$^{\text{h}}$ UT | 313 ± 30 |

### 1.3.1 Motion of the Earth through the Meteor Stream

Equation (1.15), however, computes the ideal radiant point at the time of the expected peak activity (i.e., closest approach). It does not take into account the motion of the Earth in its orbit through the entire stream. To do so requires a computation of the right ascension and declination during the complete time interval in question. Although the comet's orbit is fixed at the encounter time (i.e., all particles are assumed to be traveling parallel to the comet's orbit), the Earth is traveling through the debris ring. The relative velocity between the Earth and the particles will therefore change as a function of time, as described in Eq. (1.17):

$$\bar{v}_{\mathrm{rel}} = \bar{v}_{\mathrm{Earth}}(t) - \bar{v}_{\mathrm{comet}} \tag{1.17}$$

The Leonids are typically active for several days around the time of peak activity. To make sure the time span was not too short, however, a longer 2-week interval centered on the expected time of peak activity for the Leonids of November 17 for each year was chosen. The resultant radiant point motion is presented in Fig. 1.7. Curve-fitting results over the 2-week interval give the following equations for the mean motion of the right ascension and declination:

$$\begin{aligned} \alpha_{L\mathrm{mean}} &= 145°.84 + 0°.424{*}\mathrm{day} \\ \delta_{L\mathrm{mean}} &= 24°.67 - 0°.156{*}\mathrm{day} \end{aligned} \tag{1.18}$$

where day refers to the date in November desired (i.e., November 14, 0 hour, would compute as 14.0, while hour 12 of November 17 would be 17.5).

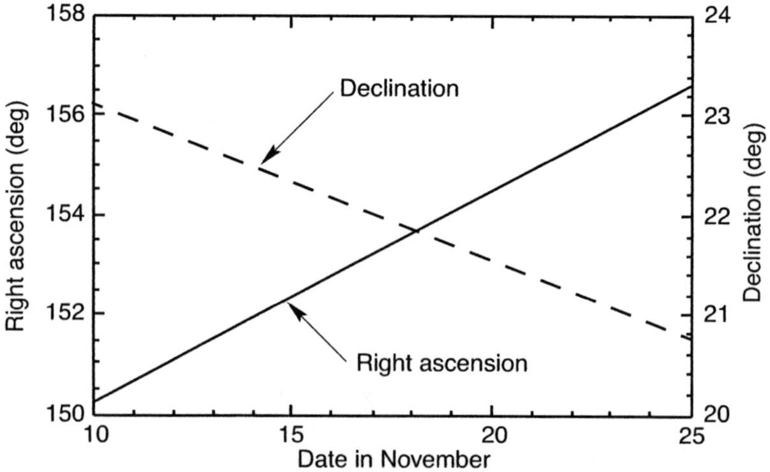


**Fig. 1.7. Leonid radiant point motion.**

### 1.3.2 Apparent Radiant Including the Motion of the Satellite

In describing the mean motion of the radiant over the course of a shower, it was assumed that the relative velocity that dictated the radiant point location was determined by the Earth and the comet (Eq. [1.14]). However, vehicles orbiting the Earth are also going to be traveling rather quickly (~7.5 km/sec for low Earth orbit [LEO]). This means that the radiant point, as far as a spacecraft is concerned, will experience a high-frequency, once-per-orbit variation about the mean values just presented. The complete relative velocity is more properly given by:

$$\bar{v}_{\text{rel}} = (\bar{v}_{\text{Earth}} + \bar{v}_s) - \bar{v}_{\text{comet}} \tag{1.19}$$

where $v_s$ is the spacecraft geocentric velocity and can be written in terms of the orbital elements for near-circular orbits as:

$$\bar{v}_s = v_s \begin{bmatrix} -\sin u \cos\Omega - \cos u \sin\Omega \cos i \\ -\sin u \sin\Omega + \cos u \cos\Omega \cos i \\ \cos u \sin i \end{bmatrix} \tag{1.20}$$

where $v_s$ is the circular velocity of the satellite, $a$ is the semimajor axis, and $\mu$ is the gravitational parameter for the Earth:

$$v_s = \sqrt{\frac{\mu}{a}} \tag{1.21}$$

and where $\Omega$, $i$, and $u$ are the right ascension of ascending node, inclination, and argument of latitude (Fig. 1.8).

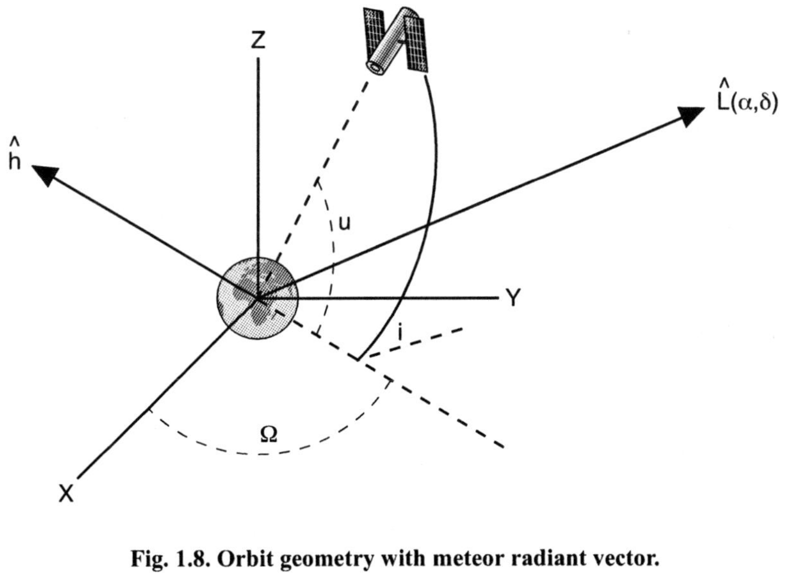


**Fig. 1.8. Orbit geometry with meteor radiant vector.**

The radiant point including the satellite's motion (where $v_{rel}$ is the relative velocity between the Earth and the comet at closest approach, Eq. [1.14]) is:

$$\alpha_L = \tan^{-1}\left(\frac{v_{\text{rel y}} - v_s \sin u \sin\Omega + v_s \cos u \cos\Omega \cos i}{v_{\text{rel x}} - v_s \sin u \cos\Omega - v_s \cos u \sin\Omega \cos i}\right)$$

$$\delta_L = \sin^{-1}\left(\frac{v_{\text{rel z}} + v_s \cos u \sin i}{[V_\infty^2 + 2(\bar{v}_{\text{rel}} \cdot \bar{v}_s) + v_s^2\,]^{1/2}}\right) \tag{1.22}$$

These expressions are complex functions of several variables:

- The approach velocity of the particles $V_\infty$
- The spacecraft velocity (i.e., the semimajor axis)
- The in-plane location $u$
- The orientation of the orbit plane as determined by the node and inclination, $\Omega$ and $i$

To simplify the problem somewhat, it is appropriate to examine the maximum envelope of values for the deviations from the mean values. Assume an orbit perpendicular to the oncoming stream ($i$ = 90 deg − $\delta_{L\text{mean}}$, $\Omega = \alpha_{L\text{mean}}$ + 90 deg). In this case, relative velocity between the particles and the spacecraft will be at its largest angular extent, subsequently giving the widest range of the radiant point locations possible (Fig. 1.9). In essence, an envelope of maximum angular values is created for any given altitude. Using the small-angle approximation, the limiting values of the maximum angular deviation are:

$$\Delta\alpha_L = \alpha_L - \alpha_{L\text{mean}} = \pm\frac{v_s}{V_\infty \cos\delta_{L\text{mean}}}$$

$$\Delta\delta_L = \delta_L - \delta_{L\text{mean}} = \pm\frac{v_s}{V_\infty}\left(\frac{1 + \sin\delta_{L\text{mean}}}{\cos\delta_{L\text{mean}}}\right) \tag{1.23}$$

Note that the plus/minus signs are independent of each other. Consider an orbit that is edge-on with respect to the meteors (Fig. 1.9, in which the viewer is

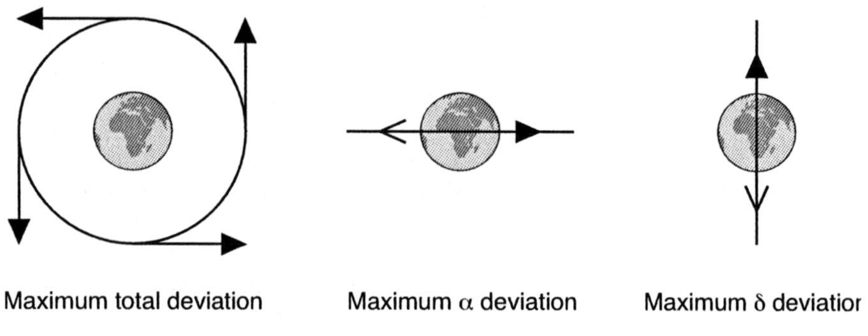


**Fig. 1.9. Radiant point deviation geometry.**

looking along the meteor approach vector). In this case, the deviation can occur either in the right ascension or declination, but not both. If the orbit is perpendicular to the approach vector, then the maximum deviation can happen in both components. The two components can therefore be independent in their behavior but are not necessarily so. Care must therefore be taken in using Eq. (1.23) to find the possible deviations. This simply determines an approximate envelope; the actual behavior will depend upon the orbit's specific orientation in space (Eq. [1.22]).

Figure 1.10 shows the variations of the maximum deviations for the Leonid meteor shower, assuming that the particular orbit of interest occurs at the time of the storm ($\alpha_{Lmean}$ = 153.45 deg, $\delta_{Lmean}$ = 21.87 deg). For LEO, the deviations from the mean can be quite large; even for geosynchronous (GEO) orbits, the values are still significant. However, this represents the maximum values possible for the Leonids. Again, each specific orbit should be examined distinctly. Each meteor event will also produce different radiant point notions. The Leonid deviations shown here can be small when compared with slower moving streams such as the Draconids.

Consider two common examples: a near-equatorial ($i$ = 5 deg) GEO satellite and a Sun-synchronous LEO satellite ($i$ = 98.60 deg, $\Omega$ = 156.93 deg). The LEO case is special in that for the Leonids, a Sun-synchronous satellite with this node and inclination is edge-on with respect to the radiant. The GEO case is significant because there are many satellites in Earth orbit with these approximate parameters.

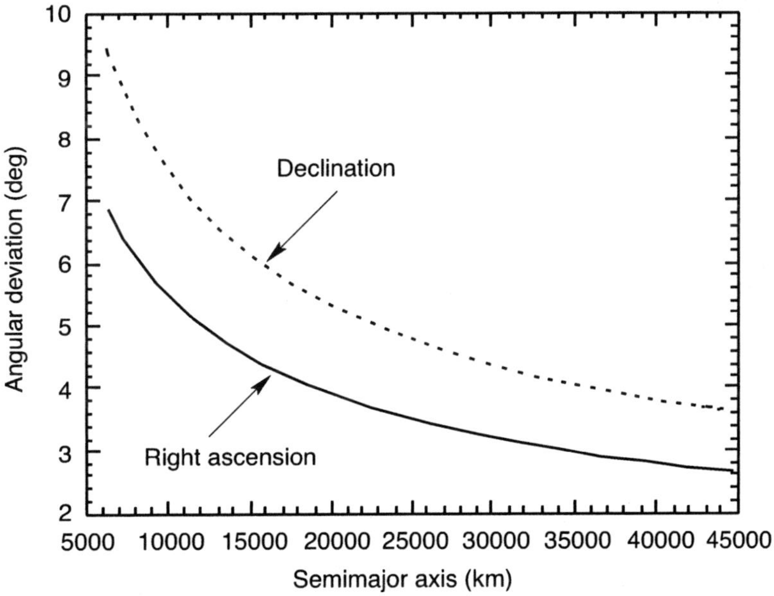


**Fig. 1.10. Maximum angular deviation as function of altitude for near-circular orbits for the Leonids.**

Figure 1.11 shows the deviation of the radiant from the mean value for these two cases over the course of an orbital cycle. For comparison purposes, note that the maximum values of the deviations for the same altitude orbit as shown in Fig. 1.10 are 6.4 deg and 8.7 deg in the right ascension and declination for the LEO altitude and 2.7 deg and 3.7 deg for GEO. From Fig. 1.11, the specific case maximums are 1.1 deg and 6.3 deg for the LEO, and 2.6 deg and 1.0 deg for GEO in the right ascension and declination, respectively. For the Sun-synchronous LEO orbit, the deviation in the declination can be almost as large as the maximum possible; since the orbit is edge-on, the right ascension variation is only about a degree from the mean value. For GEO, the declination deviation is less than a degree, while the right ascension can vary noticeably from the mean value. The larger implications of this once-per-revolution apparent motion in the radiant will be examined in Sec. 4.3.

## 1.4 Particle Flux

With the dynamical behavior of the particles determined at Earth encounter, the next issue to ponder is just how many particles have the potential of impacting an orbiting vehicle and causing damage. The temporal variation in the meteor event as viewed by the Earth is examined in terms of the activity curve as well as the distribution of the particle flux expected at encounter. This leads to a computation of the probability of impact for meteoroids likely to consist of a threat to a satellite.

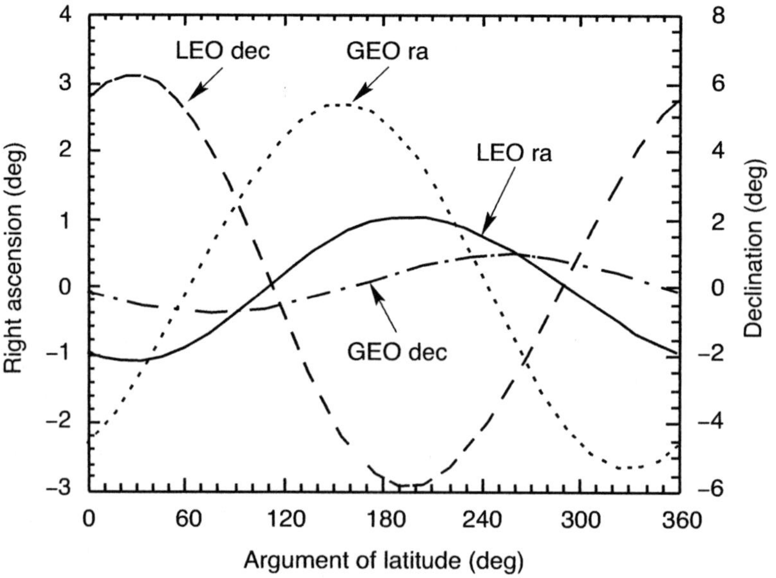


**Fig. 1.11. Radiant point motion with once-per-rev effect.**

### 1.4.1 Meteoroid Activity Curve

The Zenith Hourly Rate (ZHR) is the typical means by which the intensity of a meteor event is characterized and describes the number of meteors seen by the average observer in 1 hour under ideal sky conditions with the radiant in the zenith direction. It is most often computed from naked-eye Earth observations that have been normalized to account for atmospheric conditions and observer bias. Measurements of the meteor shower activity can also be derived from radar and infrared instruments. However, this is not done as often as naked-eye observations because of cost, availability, and the difficulty such instruments have in distinguishing background sporadic meteors and those associated with a given shower. So even with today's technology, many observations of meteor shower activity are performed by amateurs and professionals alike using visual techniques.

The observer biases inherent to naked-eye techniques are affected by the quality of the watcher's eyesight, attention span, style of observations, and judgment of meteor brightness, speed, and direction. This makes the counting of meteors highly individualistic. It is only through repetitive comparisons between observers that an individual's personal corrective factor can be determined and the observer's own biases removed from the raw data. The basic equation governing the normalization procedure (Jenniskens, 1994) is

$$\mathrm{ZHR} = \frac{c_p N}{\Delta t} r^{6.5 - L_m} \sin(h_r)^{-\gamma} \tag{1.24}$$

where $N$ is the raw observational count in the time interval $\Delta t$ (it is typically assumed that $\Delta t$ must be greater than 0.4 hour with a level of observational downtime less than 20%), $c_p$ is the observational bias factor (standard observer = 1.0), $L_m$ is the sky limiting magnitude ($L_m$ = 6.5 is ideal), $h_r$ is the radiant altitude of the particular stream (90 deg = zenith), $\gamma$ is an exponent greater than or equal to 1, but is usually taken as unity for visual observations, and $r$ is the population index determined from magnitude observations. There is a slight variation in $r$ as a function of the solar longitude (Brown and Rendtel, 1996). In addition, there is an argument for an observer dependence as well (Jenniskens, 1994). However, for the first approximation purposes a satellite operator would be interested in, it can be assumed that $r$ is constant.

Because the Earth is moving through the meteor stream for any given event, there will be a time-varying component to the measured ZHR. This activity curve defines the level of ZHR activity as a function of time for a given meteor shower (Jenniskens, 1994) as:

$$\mathrm{ZHR} = \mathrm{ZHR}_{\max} 10^{-B|\lambda - \lambda_{\max}|} \tag{1.25}$$

where $\mathrm{ZHR}_{\max}$ is the maximum zenith hourly rate for the shower under consideration, $B$ is the slope of a straight line in a log-normal plot experimentally fitted from observational data, $\lambda$ is the solar longitude, and $\lambda_{\max}$ is the solar longitude at the peak value of the ZHR. Appendix B gives the activity curve data for a variety of meteor showers.

For most meteor showers, there is an exponential growth to a sharp peak and then a fairly symmetrical shrinkage as the Earth moves away from the cometary debris. However, some of the larger meteor showers (Boo, PAu, Per, Leo, Gem, Urs) are more accurately described as the sum of two exponentials. This is largely due to the greater densities of these showers and the subsequent better observations made of them. A more definitive structure can then be determined. In an opposite sense, some showers are so sparse that modeling an exponential fit to them is problematic at best. In these cases, it is best to assume a constant ZHR over the shower's duration (aHy, dVe, tCe, Vir, CAu, eEr, dEr). For a more detailed treatment of the reduction of real observations to the ZHR activity curve, see additional sources (Jenniskens, 1994; Brown and Rendtel, 1996; and Koschack and Rendtel, 1990a and 1990b). Figure 1.12 shows a representative activity curve for the typical annual components of the Leonid shower as given by the values of Appendix B.

The length of time that the Earth spends in the stream can be approximated by integrating the time-varying component of the activity curve (Eq. [1.25]) over the entire cycle of the solar longitude:

$$\Delta t_{\text{stream}} = \frac{1}{ln 10}\sum_i \frac{1}{B_i} \tag{1.26}$$

where Δt is in the same units as *B* (i.e., per degree, per day; Appendix B gives the *B* values per degree). The index *i* ranges over the ascending and descending branches. In the case of showers such as the Leonids, where the activity curve is more accurately described by two curves (a peak and a background), the index will range over four values.

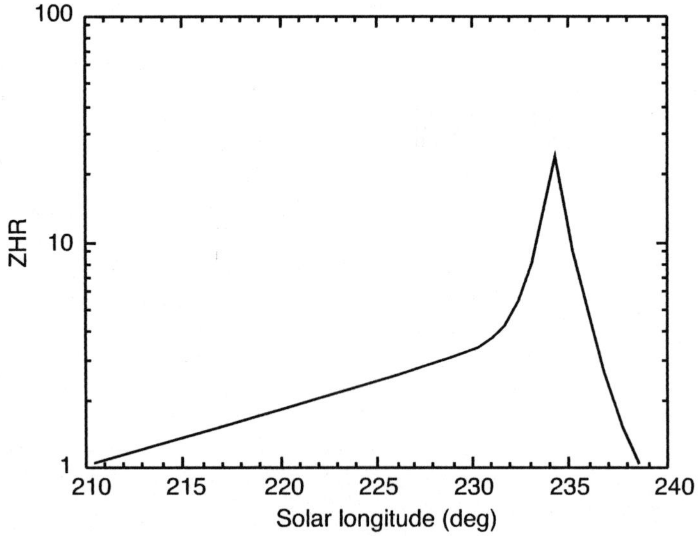


**Fig. 1.12. Typical activity curve for Leonid annual shower.**

Some care must additionally be used in defining the stream boundaries when using this expression. For the Leonids of Fig. 1.12, this expression would give an approximate length of the shower as 30 days. However, much of the activity leading up to the maximum is very sparse in nature and almost lost in the background-sporadic meteors unless the observer knows what to look for. Several of the more well-defined meteor showers show a similar characteristic. This is most likely due to the Poynting-Robertson drag pulling smaller particles in toward the Sun, making the side of the stream facing the Sun more diffuse than the side facing away. In this case it is more appropriate to assume the activity curve is symmetrical with the more well-defined branch taking the place of the ill-defined value of $B$. For the Leonids (with two components to the activity curve), Eq. (1.26) becomes:

$$\Delta t_{L-\text{shower}} = \frac{1}{ln 10}\left(\frac{2}{B^{p}} + \frac{2}{B^{b-}}\right) \tag{1.27}$$

where $p$ refers to the $B$ value for the peak, $b$ for the background component, and the minus sign indicates the descending branch. Using the Leonid values found in Appendix B, the time spent in the stream becomes 7.5 days. Assuming the Earth travels in a straight line over the course of the shower at 30 km/sec, the Earth will travel a distance in the Leonid stream tube of 19.4 million km. From the geometry of the encounter, the corresponding tube width can be found to be approximately 3 million km (the Earth goes almost directly down the tube).

For the outbursts, a similar adjustment may apply, since these events have historically been ill-defined. Appendix C gives the outburst activity curves and, assuming the 1998–2000 Leonid events will be similar to past occurrences (and assuming symmetry in the activity curve), the value of $B$ for the peak will be about 30, while the background component will have a $B$ of approximately 4. This gives the time spent in the region of greater outburst activity to be ~6 hours, with the peak region occupying less than 1 hour.

### 1.4.2 Distribution

The ZHR activity curve is thus a normalized ground-based observational count of those meteoroids that intercept the Earth's atmosphere and create a visible train large enough for the naked eye to see. So, while the ZHR activity curve gives a relative temporal measure of the meteoroid density, it does not reflect the actual particle flux that a spacecraft will experience away from the protective envelope of the atmosphere. Because the Earth observes a single pass through a meteor stream only once a year, observational confirmation of the flux is very difficult to determine. As a consequence, one of the most important assumptions in meteor stream flux modeling has been to assume that the dust in a meteor stream follows the same general type of distribution as the solar system ambient dust flux.

Although modifications have been proposed (Taylor and McBride, 1997; Wasbauer *et al.*, 1997), the interplanetary dust distribution mass-flux model most widely in use is Grün's model (Grün *et al.*, 1985):

$$\Phi_{dust}(@1\ \mathrm{AU}) = (2.20 \times 10^3 m_p^{0.306} + 15)^{-4.38} + 1.3 \times 10^{-9}(m_p + 10^{11} m_p^2 + 10^{27} m_p^4)^{-0.36} + 1.3 \times 10^{-16}(m_p + 10^6 m_p^2)^{-0.85} \quad (1.28)$$

where $m_p$ is the mass of the particle. There are three terms in this model: the first line is dominant for larger particles ($m > 10^{-9}$ g) that pose the greatest threat to spacecraft; the second term is more significant for medium-sized particles ($10^{-14}$ g $< m < 10^{-9}$ g); while the third describes the smaller dust motes ($m < 10^{-14}$ g).

For penetration to occur, only those particles that have a mass greater than $10^{-7}$ g need to be considered (see Sec. 4.2.1 for the ballistic limiting equations that assume the worst case of a single wall of aluminum for the fast-moving Leonids); the Grün model can be simplified to:

$$\Phi = 2.292 \times 10^{-15} m_p^{-1.34} \quad (1.29)$$

But while this describes the ambient solar system flux of the dust for threatening-sized particles, it does not describe a particular meteor stream's flux distribution. Since smaller particles resident in the solar system will be "cleaned up" by Poynting-Robertson drag, it is logical to assume that the meteor stream mass index (exponent in Eq.[1.29]) will be different from the solar system dust mass index. The more general form of the instantaneous meteor stream flux at a particular mass value is:

$$\Phi = N_0 V_\infty m^{-s} \quad (1.30)$$

where $s$ is referred to as the mass index for the stream and $N_0$ is a constant particular to that stream. The mass index is related to the population index $r$ that was mentioned in the previous section as (McKinley, 1961):

$$s = 1 + 2.5 \log(r) \quad (1.31)$$

To find the total number of particles a spacecraft will face, this expression must be integrated from an arbitrary small mass to infinity:

$$\Phi_T = \frac{N_0 V_\infty}{s-1} m^{1-s} \quad (1.32)$$

But $N_0$ is not a known value and must be solved for. The only parameter that is available to accomplish this comes from the observations inherent in the ZHR activity curve. The expression relating the ZHR to the spatial number density (Brown and Rendtel, 1996) is:

$$N_{+6.5} = \frac{\mathrm{ZHR}_{obs} C'}{3600 A_{sky} V_\infty} \quad (1.33)$$

where the typical units for this expression in the literature is $10^{-9}$ km$^{-3}$. Since the ZHR is normalized about a magnitude of +6.5, $N_{+6.5}$ is the spatial number density for that magnitude only.

The factor $C'$ adjusts the observed ZHR to a true value accounting for meteors missed by observers. $C'$ is a function of the population index (Koschack and Rendtel, 1990b):

$$C' = 13.10r - 16.45 \tag{1.34}$$

The parameter $A_{sky}$ is the effective collecting area for the meteors in the atmosphere at an assumed height of 100 km that an observer is able to watch. As such, $A_{sky}$ is also a function of the population index (Koschack and Rendtel, 1990a):

$$A_{sky} = 37200(r - 1.3)^{-0.748}\ (\text{km}^2) \tag{1.35}$$

The flux at magnitude +6.5 is simply the spatial number density at that magnitude multiplied by the approach velocity:

$$\Phi_{+6.5} = N_{+6.5}V_{\infty} \tag{1.36}$$

Once the magnitude +6.5 flux is known, it is a simple matter to scale it for different masses as well as the presence of an outburst:

$$\Phi_T = \left(\frac{\text{ZHR}_{\text{outburst}}}{\text{ZHR}_{\text{shower}}}\right) F_s \left(\frac{m}{m_{+6.5}}\right)^{1-s} \Phi_{+6.5} \tag{1.37}$$

where $F_s$ is a mass index enhancement factor that describes the greater concentration of smaller particles that would occur in an outburst as compared with the normal shower. In essence, the Poynting-Robertson drag has not had a chance to remove the smaller particles from the system for the near-comet types of outbursts. $F_s$ can be approximated by Table 1.6 (Beech *et al.*, 1997).

The magnitude +6.5 particle mass in Eq. (1.37) is found from the mass-luminosity relation:

$$L_m = A\log(V_{\infty}) + B\log(m_{+6.5}) + C \tag{1.38}$$

where $L_m$ is the visual magnitude of the streak (i.e., $L_m$ = +6.5), $V_{\infty}$ is the approach velocity of the meteoroid, and $m_{+6.5}$ is the mass of the particle. The constants A, B, and C vary between models with three examples given in Table 1.7 (Cooke and Anderson, 1998).

Also shown in Table 1.7 are the masses in grams of the ambient and Leonid ($V_{\infty}$ = 71 km/sec) particles for the ideal limiting magnitude. Nearly an order of magnitude difference in the resulting meteoroid flux can therefore result from the various models. Analysts should note that the flux modeling can be off by a similar order of magnitude as a consequence of these uncertainties.

**Table 1.6. Storm Enhancement Factor**

| | | | | | | |
|---|---|---|---|---|---|---|
| s | 1.5 | 1.75 | 2.0 | 2.25 | 2.5 | 2.75 |
| F | 0.1 | 0.4 | 1.0 | 2.0 | 3.8 | 6.5 |

**Table 1.7. Meteor Mass-Luminosity Model Parameters**

| A | B | C | Author | Ambient | Leonid |
|---|---|---|---|---|---|
| –6.27 | –1.61 | 9.77 | Jenniskens | 1.4e-3 | 6.6e-6 |
| –8.75 | –2.25 | 11.59 | Brown | 2.4e-3 | 1.2e-5 |
| –8.75 | –2.25 | 12.90 | Kessler | 9.2e-3 | 4.4e-5 |

By combining Eqs. (1.33) through (1.37), the general form for the equation for the cumulative flux of particles having a mass greater than a specified value $m$ for any given meteor event becomes (particles/$m^2$–$s$):

$$\Phi_T = k_m \mathrm{ZHR}\ m^{1-s} \tag{1.39}$$

where $k_m$ is a constant that is event specific:

$$k_m = 7.467 \times 10^{-15} F_s (m_{+6.5})^{s-1} (13.10r - 16.45)(r - 1.3)^{0.748} \tag{1.40}$$

Evaluating Eq. (1.39) for the Leonids (assuming a mass index of 2.2 for the outbursts and Brown's value of the $m_{+6.5}$ particle as the middling value in Table 1.7) gives the final expression for the cumulative flux of particles for the 1998–2000 expected outbursts:

$$\Phi_L = 6.3 \times 10^{-19} \mathrm{ZHR}_{\mathrm{outburst}} m^{-1.2} \qquad (\text{particles}/m^2 - s) \tag{1.41}$$

In comparison, the annual Leonid shower has a flux of (mass index of 2.0):

$$\Phi_L = 1.7 \times 10^{-18} \mathrm{ZHR}_{\mathrm{shower}} m^{-1.0} \qquad (\text{particles}/m^2 - s) \tag{1.42}$$

### 1.4.3 Impact

Damage from a particle impact can come from a variety of phenomena either mechanical (shock waves, cratering, penetration, spall production) or electrical (plasma production leading to electrostatic discharge [ESD], electromagnetic impulse). This study is not designed to address the physics of hypervelocity phenomenology but rather the dynamics of mitigating the threat that they pose. As a consequence, the interest here is to determine the minimum-mass particle that may cause significant levels of danger. This minimum mass provides the basis for determining the flux that has the potential to hurt the vehicle and therefore the probability of a dangerous impact. Assuming the worst case of a single wall of thin 1-mm aluminum and a meteoroid particle impacting perpendicular to the spacecraft surface, the ballistic limiting equation (Sec. 4.2) reduces to a relation between the mass of the particle and its approach velocity:

$$m_c \sim \frac{7.3 \times 10^{-3}}{V_\infty^2} \tag{1.43}$$

where $V_\infty$ is in km/sec and the mass is in grams. This gives an approximate measure of the minimum-size particle that can be a cause for worry.

The particles that hit a spacecraft of a given cross-sectional area over a meteor event of known duration can be determined from the flux as:

$$N = \int_{t_1}^{t_2} \Phi_T A_{s/c} dt \tag{1.44}$$

where $\Phi_T$ is the flux (#/m$^2$/s), $A_{s/c}$ is the cross-sectional area exposed to the meteors (m$^2$), and the meteor event exists over the time span from $t_1$ to $t_2$ (sec). Figure 1.13 shows this expression for both the Leonid shower and the expected worst-case outbursts of 1998–2000 (Appendix B data were used for the Leonid activity curve with a ZHR of 10,000). Since the ZHR is time-varying, Eq. (1.44) is computed in 1-hour intervals over which the ZHR is assumed constant for the shower, and in 10-minute intervals over the course of the expected storm. The number of impacts is also normalized with respect to the spacecraft area in order to keep the results general.

The impact probabilities are therefore potentially high during the time of the storm. Note that this is a time profile determined in 10-minute sections. To find the total probability of a hit, the number of impacts under the outburst curve must be totaled. Doing so yields a 0.002 probability. The small time of peak activity clearly dominates the result. A satellite with a 100-m$^2$ area would thus have had roughly a 2-in-10 chance of being hit by a dangerously sized particle during the 1998–2000 Leonids, assuming a peak worst-case ZHR of 10,000. With more than 500 U.S. and international satellites currently in operation, and assuming an average area of 10 m$^2$, approximately five satellites are going to be

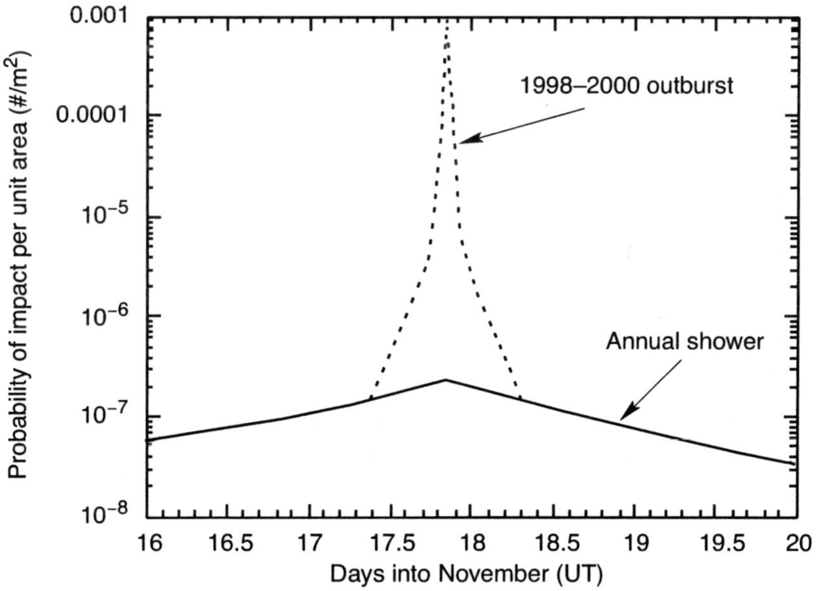


**Fig. 1.13. Impact probabilities for critical mass particles.**

impacted by a dangerous mass particle during the 1998–2000 event. However, it must be stressed again that the flux estimates could be off by an order of magnitude, either in a positive (i.e., impact chances are lower) or negative (chances become close to unity for a 100 $m^2$ satellite) way.

Figure 1.14 shows the Leonid flux per unit area at the time of peak activity as a function of the particle mass for both the annual shower and the expected 1998–2000 outburst. The smaller particles pose the greatest danger to spacecraft; their presence is orders of magnitude more prevalent than the larger meteoroids. In comparison, launch and mission operations become concerned about the possibility of collision with orbital debris when the probabilities are on the order of $10^{-6}$. This level of probability would roughly correspond to a Leonid particle that would have the same kinetic energy as a 0.22 caliber bullet. Clearly, the Leonid outburst will present a danger much larger than typically accepted for mission operations. However, the danger will come from the small-sized particles that cause plasma and perforation damage, which could render the satellite inoperable. Catastrophic destruction of the vehicle is not likely.

## 1.5 Summary of Threat Assessment

A summary of the relevant procedures necessary to assess a meteoric threat is presented in Algorithm 1.1. A summary of the 1998–2000 Leonid encounter data is given in Appendix D.

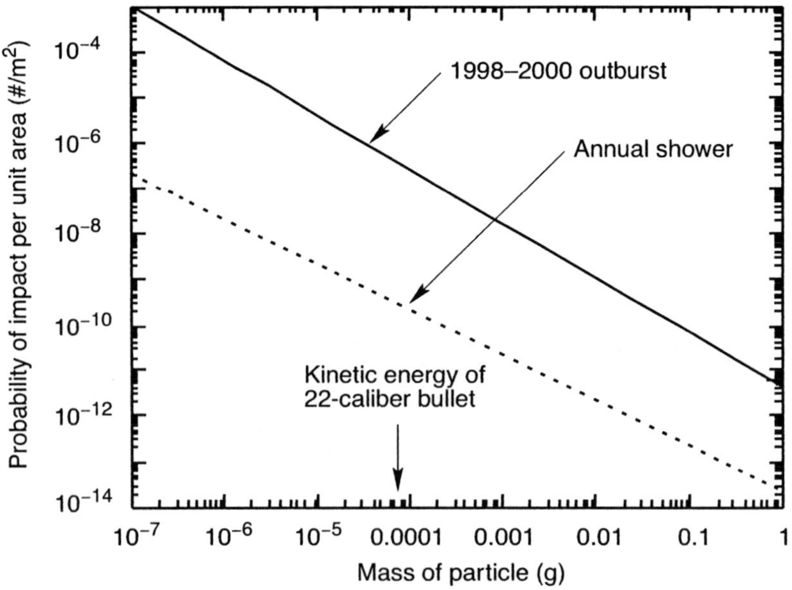


**Fig. 1.14. Impact probabilities at Leonid peak**

1. Compute the parent comet's orbit.

$$\frac{d^2\bar{r}}{dt^2} = -\mu_{\text{sun}}\frac{\bar{r}}{r^3} + \sum_{j=1}^{N}\mu_j\left[\frac{\bar{r}_j-\bar{r}}{\|\bar{r}_j-\bar{r}\|^3} - \frac{\bar{r}_j}{\|\bar{r}_j\|^3}\right] + g(r')[A_1\hat{R} + A_2\hat{T}]$$

$$+\frac{\mu_{\text{sun}}}{r^3}\left\{\left[\frac{2(\gamma+\beta)}{c^2}\frac{\mu_{\text{sun}}}{r} - \frac{\gamma}{c_2}v^2\right]\bar{r} + \left[\frac{2(\gamma+1)}{c^2}(\bar{r}\cdot\bar{v})\right]\bar{v}\right\}$$

2. For meteor system evolution, replace the rocket force, $g$, with the true nonconservative forces:

$$\bar{F}_{\text{noncons}} = \beta\frac{\mu_{\text{sun}}}{r^2}\left[\left(1-\frac{v_R}{c}\right)\frac{\bar{r}}{r} - \left(\frac{1+sw}{c}\right)\bar{v}\right]$$

Add in random velocities near perihelion to simulate the particle cloud evolution:

$$v_{\text{eject}} = \left(\frac{4R_c}{n_{\text{eff}}s_p r^{9/4}} - 0.052R_c^2\right)^{1/2} \times 0.00328 \text{ km/s}$$

3. Compute the once-per-rev radiant point from the relative velocity at point of closest approach:

$$\alpha_L = \tan^{-1}\left(\frac{v_{\text{rel }y} - v_s\sin u\sin\Omega + v_s\cos u\cos\Omega\cos i}{v_{\text{rel }x} - v_s\sin u\cos\Omega - v_s\cos u\sin\Omega\cos i}\right)$$

$$\delta_L = \sin^{-1}\left(\frac{v_{\text{rel }z} + v_s\cos u\sin i}{[V_\infty^2 + 2(v_{\text{rel}}\cdot v_s) + v_s^2]^{1/2}}\right)$$

4. The critical mass to be considered dangerous for spall creation is approximately ($m$ in g, $V_\infty$ in km/sec):

$$m_c \sim \frac{7.3\times10^{-3}}{V_\infty^2}$$

Electrostatic discharge is difficult to quantify since any charge is potentially dangerous depending upon the latent charge already on the spacecraft.

5. Compute the particle flux for all particles of this size mass and larger:

$$\Phi_T = k_m \text{ZHR } m_c{}^{1-s}$$

6. The number of impacts for a short-duration outburst become:

$$N = \Phi_T A_{s/c}\Delta t_{\text{event}}$$

**Algorithm 1.1. Meteoric outburst threat assessment.**

# 2 The Geometry of Shadows: Meteoric Eclipses

When meteor streams encounter the near-Earth environment, serious problems can result for satellites that are orbiting the Earth. If the meteor stream is dense enough, loss of the vehicle or of the mission can result from either a physically damaging impact or an electrostatic discharge. One way to mitigate the probability of a strike is for the spacecraft to take cover behind a larger planetary body. This chapter examines the geometry of meteoric eclipses, which occur when the satellite experiences an eclipse with respect to the oncoming meteors (this should not be confused with lunar or solar eclipses), the optimal orbit configuration for an eclipse, and the limitations on the effectiveness of the Earth's shadow in minimizing the potential for a meteoroid impact. A detailed analysis of the deltaV requirements to move a given satellite from its current location into the optimal configuration will be conducted in Chapter 4.

## 2.1 Earth Shadowing

Although some satellites can be considered to operate solely in the lunar environment (Clementine, Lunar Prospector), the vast majority of spacecraft in the Earth-Moon system orbit close to the Earth. Consequently, the Earth is the body that has the potential of providing the greatest amount of shadowing protection. However, this protection is limited to that portion of the satellite's orbit that is behind the Earth with respect to the oncoming meteoroids. The benefit gained is thus time dependent; when the vehicle is in the shadow cast by the Earth, it is completely protected, and a zero probability of impact results; when the vehicle is not in the shadow, the probability of impact is not mitigated at all (Fig. 2.1). The central idea, then, is not only to orient the orbit plane to take advantage of the shadowing, but to place the satellite such that it spends as much time as possible behind the Earth. While the analysis is oriented specifically toward the Earth and the Leonid showers, the equations are left in general form in order to be

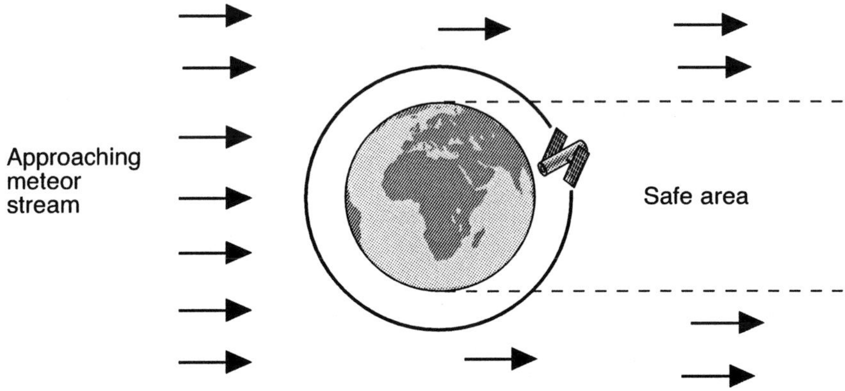


**Fig. 2.1. Meteoric eclipse cast by the Earth.**

applicable to any orbit about any central body for any known meteor event. Either way, there is no completely safe location for spacecraft save one: being behind a body such as the Earth or Moon when a short-duration storm occurs (Fig. 2.1).

### 2.1.1 Earth Shadowing for Near-Circular Orbits

From a simple geometric examination of the problem, it can be seen that orbits that traverse directly behind the Earth with respect to the incoming meteor stream will be the most protected (Fig. 2.2). When the orbit is in this orientation, the satellite has the potential to spend the greatest amount of time in the Earth's shadow. If it is not, protection will decrease until no part of the orbit is covered.

To compute the optimal orientation for near-circular orbits (i.e., those orbits that have near-constant angular in-plane motion), it is a simple matter to take the dot product of the vector normal to the orbit plane (i.e., the unitized angular momentum vector) and the meteor radiant vector. This product should go to zero when the orbit is going through the desired *L*. In other words, the orbit plane needs to contain the meteor radiant vector. Given that while the inclination and the node define the orientation of the orbit plane, it is only the node that changes temporally because of the secular precession of the Earth's oblateness. Therefore, the problem will be to fix the inclination and solve for the optimal node location (Fig. 2.3).

Assume the Leonids are approaching the Earth with velocity $V_\infty$ from the unit direction *L*, where *L* is composed of the right ascension $\alpha$ and declination $\delta$ of the meteor shower radiant point ($\alpha = 153.45$ deg, $\delta = 21.87$ deg for the Leonids) as:

$$\hat{L} = [\cos\delta\cos\alpha,\ \cos\delta\sin\alpha,\ \sin\delta]^T \tag{2.1}$$

and where the orbit being approached has the unitized angular momentum vector written in terms of the orbit elements as:

$$\hat{h} = [\sin\Omega\sin i,\ -\cos\Omega\sin i,\ \cos i]^T \tag{2.2}$$

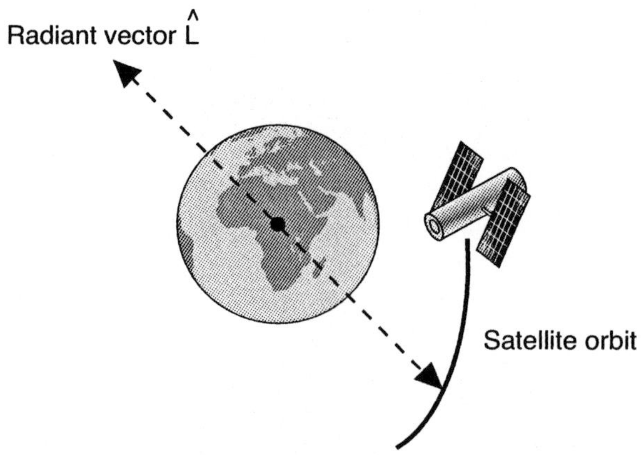


**Fig. 2.2. Meteor radiant vector and orbit geometry.**

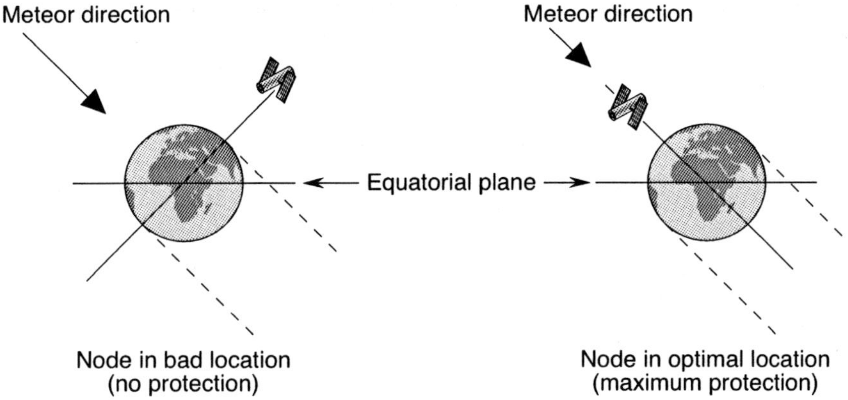


**Fig. 2.3. Protection from different node locations for same orbit inclination.**

with Ω and $i$ being the usual Keplerian orbit elements of the right ascension of the ascending node and the inclination (Fig. 2.4). Both $h$ and $L$ are given in terms of the geocentric equatorial coordinate system.

The dot product is computed as:

$$\hat{h} \cdot \hat{L} = \cos(90 - \zeta) = \cos\delta \sin i \sin(\Omega - \alpha) + \sin\delta \cos i \tag{2.3}$$

where ζ is similar to the usual beta angle used in solar eclipse studies. The variable ζ will be applied in this study to emphasize the difference between the two angles, namely that the beta angle is derived for the Sun and its eclipses, the zeta angle for the meteor showers and their eclipses.

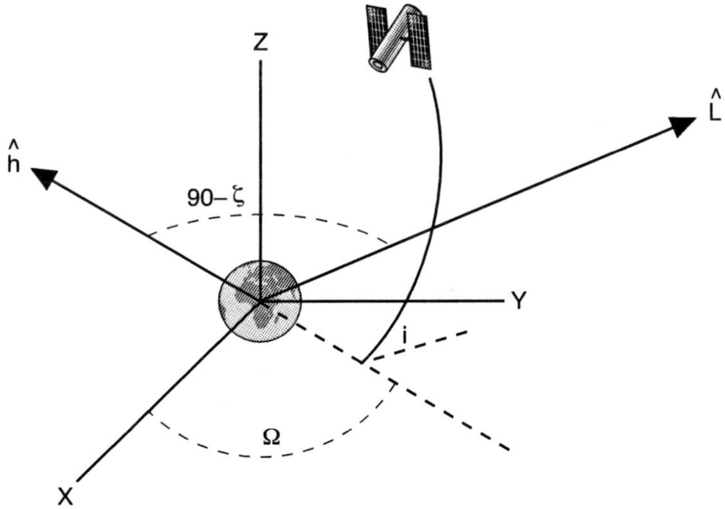


**Fig. 2.4. Orbit geometry with Leonid approach.**

When the orbit is aligned with the approach vector, the dot product goes to zero, giving the desired nodal relation:

$$\sin(\Omega - \alpha) = -\frac{\tan\delta}{\tan i} \tag{2.4}$$

The solutions for the right ascension of the ascending node are therefore:

$$\Omega = \alpha + \sin^{-1}\left(-\frac{\tan\delta}{\tan i}\right) \quad \text{and} \quad \Omega = \alpha + 180 - \sin^{-1}\left(-\frac{\tan\delta}{\tan i}\right) \tag{2.5}$$

When the satellite has an inclination lower than that of the meteor shower's declination, the above expression is unsolvable, and instead of aligning the orbit in such a way as to include the meteor radiant vector, an orbit geometry must be chosen such that the dot product minimizes instead of going to zero. This is accomplished by setting the right ascension of the ascending node to be perpendicular to the right ascension of the incoming meteors. However, these low-inclination orbits cannot provide optimal coverage and may even be so shallow and far out that no shadowing is possible at all. Such instances of partial coverage will be discussed further in the next section. For this reason, the solutions of Eq. (2.4) are termed optimal because they contain the meteor radiant vector. The resulting optimal solution can be plotted as shown in Fig. 2.5 for two relevant meteor showers: the Leonids and the Draconids.

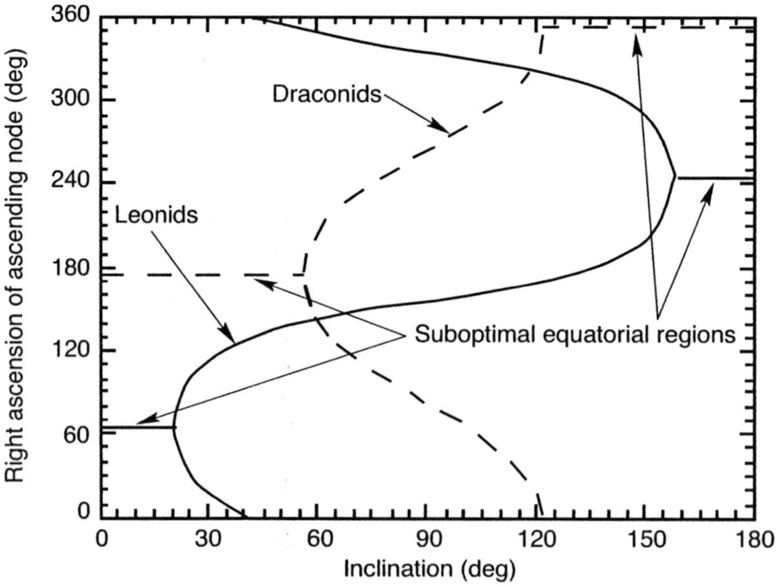


**Fig. 2.5. Right ascension of ascending node for maximal shadowing protection.**

The optimal node location for a near-circular orbit has been derived (Eq.[2.4]), but how effective is this? To answer this question, consider that the satellite is traveling at a constant angular rate (circular orbit) in the optimal geometry. Then, naturally the benefit gained through Earth shadowing is going to be a function of the time spent in the shadow. For circular orbits, once the geometry is determined, the time in umbra is a direct linear relation to the shadowed angular arc and is solely a function of the satellite's semimajor axis $a$, as shown in Fig. 2.6, where $R_e$ is the Earth's radius and $h_a$ is the height of the effective atmosphere.

The expression for the angle $\zeta^*$ that denotes one-half of the regime of protection afforded by the Earth's umbra is therefore (where the sense of the angle need not be considered since $\zeta^*$ is constrained to always be less than 90 deg):

$$\zeta^* = \sin^{-1}\left(\frac{R_e + h_a}{a}\right) \tag{2.6}$$

The actual benefit gained is reflected as that portion of the orbital period spent in the shadow as opposed to the total period, or expressed as a percentage in terms of the angle $\zeta^*$ (in degrees):

$$\%\ \text{protection} = 100.\ *\ 2\ \zeta^*/360 \tag{2.7}$$

while the total time spent in the Earth's shadow is therefore the percentage multiplied by the total period:

$$\Delta t_{\text{shadow}} = \left(\frac{2\zeta^*}{360}\right) 2\pi \sqrt{\frac{a^3}{\mu}} \tag{2.8}$$

The results of Eqs. (2.7) and (2.8) are shown in Fig. 2.7 as a function of the altitude assuming the effective atmosphere extends 100 km above the surface of a spherical Earth. As can be seen, the benefits appear to drop off significantly once the regions close to the Earth are left behind (solid curve, Eq. [2.7]). However, this can be misleading. If the timing of the approaching storm is known, the satellite can be phased into an in-plane angle to take full advantage of the shadowing (dashed curve showing the total time of protection, Eq. [2.8]) by means of

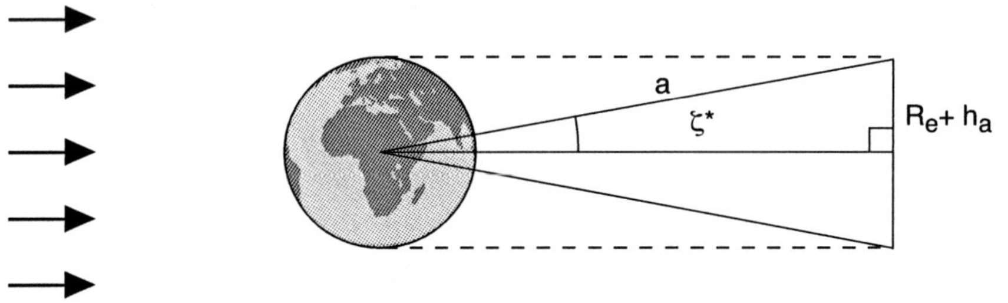


**Fig. 2.6. Protected portion of the satellite's orbit.**

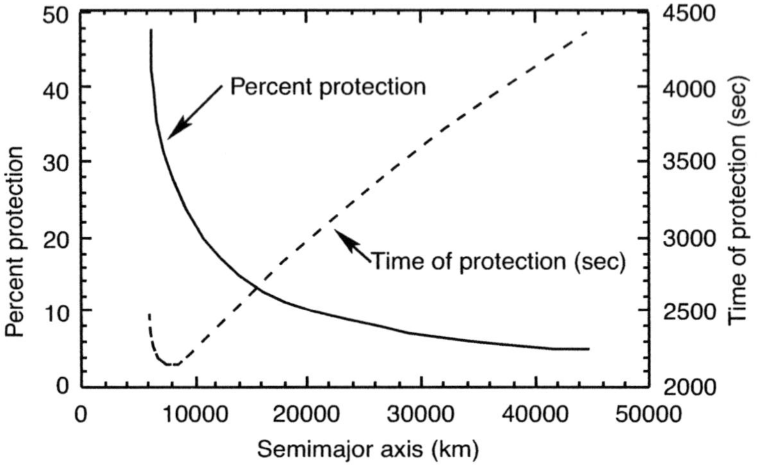


**Fig. 2.7. Shadowing effectiveness for near-circular orbits.**

adjusting the true anomaly. In this case the farther out the satellite is, the more beneficial. With geosynchronous satellites, over an hour's worth of safety is possible.

Note that in computing the percentage, no account is made for a short-duration storm. In essence, the storm is assumed to be much longer than the orbital period (i.e., a meteor shower of a week duration). Alternatively, the percentage can be thought of as the probability of shadowing for a short-duration storm. For example, assume a LEO satellite is in the midst of a long meteor event. Then from Fig. 2.7, about 50% of the time, the satellite will be in the Earth's shadow through multiple revolutions. Conversely, if the event is of short duration such as a storm, the satellite has a probability of 50% at any given time of being in the Earth's shadow when the storm hits and therefore protected.

As another example, consider a meteor storm of 2 hours duration that encounters two satellites, one at LEO with a semimajor axis of 7500 km and the other at GEO with a semimajor axis of 42,240 km. In both cases, the meteor approach vector must lie in the orbit plane. Then the lower satellite, which is whipping through its orbit with a 108-min period, will experience on average about 40% protection, meaning just over half of the meteors would impact the satellite that otherwise would occur if the Earth were a nondimensional point mass (i.e., the particles could travel through the Earth). For a geosynchronous satellite, however, the timing of the storm is critical. On average the GEO satellite would have a chance of reducing the meteor threat by only about 5%. But if the precise start time of the 2-hour storm is known, the satellite can be phased into a regime of maximum protection, affording it approximately 4200 sec of protection, implying that only about 40% of the incoming meteors would have a chance of impacting the satellite for the 2-hour storm. This is a critical point. If the timing of the storm is not known, the lower the satellite is, the better. The satellite on average will spend more of its time in the shadow. If the timing is known, higher-altitude satellites can be placed into regions of maximum protection that may (depending upon the length of the storm) provide better results than do those applicable to LEO satellites. A campaign of deliberate shadowing becomes possible in this case. A summary of the relevant equations is given in Algorithm 2.1. The algorithms presented at the end of each of these subsections summarize the derivations presented for easier reference.

1. Optimal node: $$\sin(\Omega_{opt} - \alpha) = -\frac{\tan\delta}{\tan i}$$

2. Portion of the orbit that gets shadowed: $$2\zeta^* = 2\sin^{-1}\left(\frac{R_e + h_a}{a}\right)$$

3. Time spent in the shadow: $$\Delta t_{shadow} = 2\zeta^* \sqrt{\frac{a^3}{\mu}}$$

**Algorithm 2.1. Optimal near-circular orbit shadowing equations.**

### 2.1.2 Partial Shadowing for Near-Circular Orbits

The optimal nodal solution just derived is exactly that: optimal. It is optimal in the sense that the orbit was constrained to pass through the meteor radiant vector. If it does not, benefit can still be derived from Earth shadowing but to a lesser extent than the optimal solution. For example, Fig. 2.5 shows the optimal nodal solution but, as has been mentioned, no such optimal solution exists for near-equatorial orbits (those orbits that have an inclination shallower than the declination of the approaching meteors). For these orbits, only partial coverage is possible. The criteria to establish whether the satellite passes into the Earth's meteoric shadow at some point in its orbit at all can be found from simple geometry. The angular distance from the approach vector multiplied by the radius must be less than the cylindrical extent of the shadow:

$$a|\hat{L} \cdot \hat{h}| \leq R_e + h_a \tag{2.9}$$

If this condition is satisfied, some shadowing occurs. In terms of the orbit elements, the dot product is found from Eq. (2.3); therefore Eq. (2.9) reduces to

$$-\frac{R_e + h_a}{a} \leq \cos\delta\sin i\sin(\Omega - \alpha) + \sin\delta\cos i \leq \frac{R_e + h_a}{a} \tag{2.10}$$

The range of the node that is necessary can be solved for by taking the minimum and maximum of Eq. (2.10).

$$\sin(\Omega - \alpha) = -\frac{\tan\delta}{\tan i} \pm \frac{R_e + h_a}{a\ \cos\delta\sin i} \tag{2.11}$$

The ranges are going to be dependent now not only upon the inclination of the orbit, but also upon the semimajor axis. Figure 2.8 shows the results of this calculation for Leonids, considering two satellites: one in an LEO orbit, the other in

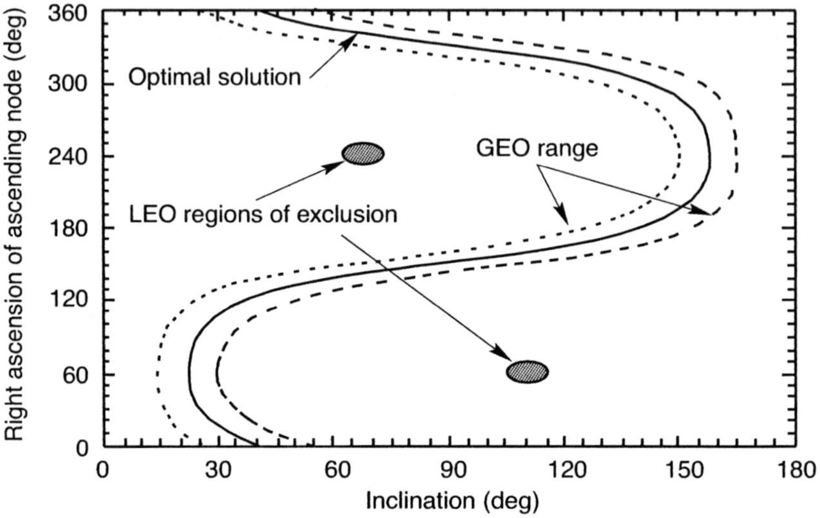


**Fig. 2.8. Leonid nodal range for suboptimal solution.**

GEO. The LEO exclusion zone represents those orientations that do *not* provide *any* coverage. In essence, the whole range of nodes and inclinations provides some semblance of shadowing except for these small regions for low-altitude vehicles. Physically, these regions represent orbits that lie nearly perpendicular to the incoming meteors. The GEO exclusion zone basically covers the whole plot except for a small region close to the optimal. The GEO range of allowable nodes is quite tight as compared with the LEO range. Because they are so far out, less deviation of the node is allowed if shadowing is to be maintained.

Even though partial coverage is considered, there will be a value for the semimajor axis such that near-equatorial orbits cannot provide any shadowing at all regardless of the nodal orientation. As the semimajor axis increases, the LEO exclusion zone will grow until the equatorial inclination is reached. Equivalently, the GEO inclusion range will grow as the semimajor axis comes in from that height until an inclination of zero is reached. The value of the semimajor axis that corresponds to this scenario can be found by letting the inclination go to zero in Eq. (2.10):

$$a \le \left| \frac{R_e + h_a}{\sin\delta} \right| \tag{2.12}$$

For equatorial near-circular orbits having a larger semimajor axis than this value, any coverage at all becomes impossible. Assuming that the shielding atmosphere extends 100 km above the Earth's surface gives a value of the critical semimajor axis for the Leonids of 17,293 km. This value assumes a zero inclination; for small values of the inclination this critical semimajor axis value will become larger. Equatorial near-circular satellites with semimajor axes greater than this value will not experience any shadowing.

Solving the reverse problem, the minimum inclination for a specific altitude satellite can be expressed as (where the inclination is assumed to be small):

$$|i \cos\delta + \sin\delta| = \frac{R_e + h_a}{a} \tag{2.13}$$

For a GEO satellite, the corresponding minimum inclination is approximately 14 deg for the Leonids. When a GEO satellite has an inclination less than this value, no Earth shadowing is possible.

Similar to the optimal case, the percent coverage can be solved for, although the problem is more complex here. The relevant value is still the angle $\zeta^*$ that the orbit traverses during the course of its time in the shadow. But the solution must be found from spherical geometry, as shown in Fig. 2.9. The angle $\zeta_{\text{optimal}}$ is that zeta angle derived for the optimal case (Fig. 2.6) and the angle $\zeta$ (Fig. 2.4) is representative of the dot product of the meteor approach vector and the unitized angular momentum vector of the orbit plane (i.e., how close to the optimal condition does the orbit plane come). The closest point of the orbit to the approach vector is produced by a right angle in a spherical sense.

The angles $\zeta$ and $\zeta_{\text{optimal}}$ are found from known quantities (Eqs. [2.3] and [2.6], respectively). Then from spherical trigonometry, the angles are related through the formula (Wertz and Larson, 1991):

$$\cos\zeta^* = \frac{\cos\zeta_{\text{optimal}}}{\cos\zeta} \tag{2.14}$$

Substituting the appropriate expressions, the angle $\zeta^*$ can be found, and Eq. (2.6) describing the arc length of the orbit shadowing then modifies in the case of the suboptimal node to:

$$\zeta^* = \sin^{-1}\left\{\left[\frac{\left(\frac{R_e + h_a}{a}\right)^2 - (\hat{L}\cdot\hat{h})^2}{1 - (\hat{L}\cdot\hat{h})^2}\right]^{1/2}\right\} \tag{2.15}$$

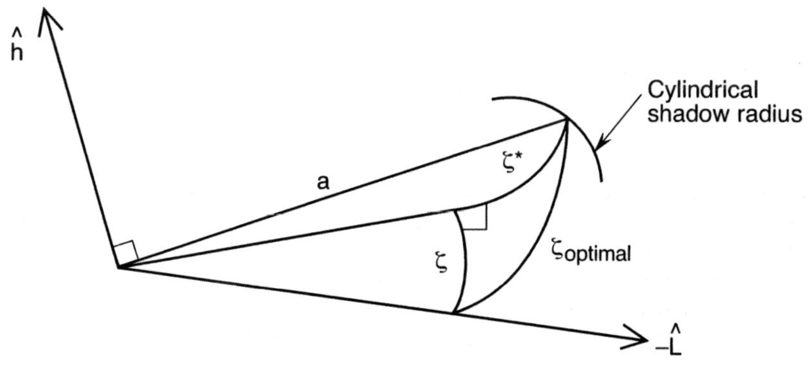


**Fig. 2.9. Orbital arc for suboptimal node.**

where the dot product in terms of the orbit elements is found from Eq. (2.3). As the orbit approaches the meteor radiant vector, this expression reduces to that of Eq. (2.6); when the condition satisfying the limits of the cylindrical shadowing (Eq. [2.9]) is considered, Eq. (2.15) gives the angle $\zeta^*$ to be appropriately equal to zero.

1. Optimal node: $$\sin(\Omega_{opt} - \alpha) = -\frac{\tan\delta}{\tan i}$$

2. Range of nodes: $$\sin(\Omega - \alpha) - \sin(\Omega_{opt} - \alpha) \le \left|\frac{R_e + h_a}{a \;\; \cos\delta \sin i}\right|$$

3. Portion of the orbit that gets shadowed:

$$2\zeta^* = 2\sin^{-1}\left\{\left[\frac{\dfrac{R_e + h_a{}^2}{a} - (\hat{L}\cdot\hat{h})^2}{1 - (\hat{L}\cdot\hat{h})^2}\right]^{1/2}\right\}$$

4. Time spent in the shadow: $$\Delta t_{shadow} = 2\zeta^*\sqrt{\frac{a^3}{\mu}}$$

**Algorithm 2.2. Suboptimal near-circular orbit shadowing equations.**

### 2.1.3 Earth Shadowing for Eccentric Orbits

No appreciable benefit exists through additional adjustment of the orbit geometry for near-circular orbits. These orbits have near-constant angular rates, so in-plane orientation changes (i.e., altering the perigee location) do not affect the solution. However, for orbits with a significant value of the eccentricity, this is not the case. Eccentric orbits do have significant changes in the angular rate at which the satellite is moving with respect to the central body, and this can be taken advantage of. Because the satellite must be behind the Earth in order to be favorably protected by shadowing effects, placing the "slow" portion of the orbit behind the Earth and leaving the satellite exposed during the "fast" phase of its motion, Earth shadowing can be maximized. This corresponds to aligning the meteor approach vector with the perigee vector of the orbit. Because the satellite moves fastest at perigee and slowest at apogee, this in-plane orientation will optimize the shielding afforded by the Earth (Fig. 2.10).

The near-circular solution for the optimal node continues to be valid for the eccentric orbit, as the unitized angular momentum vector does not depend upon the eccentricity of the orbit. The dot product of the unitized angular momentum vector and the approach vector goes to zero under the same circumstances, regardless of the orbit shape. Therefore, the optimal nodal solution (Eq. [2.4]) is still valid for elliptical orbits (Eq. [2.16]):

$$\sin(\Omega - \alpha) = -\frac{\tan\delta}{\tan i} \tag{2.16}$$

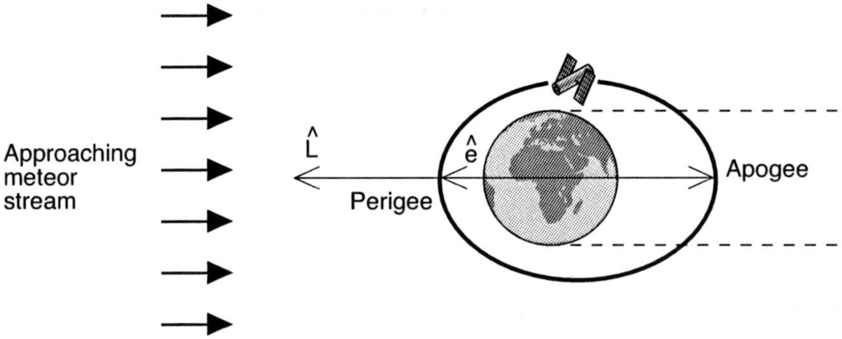


**Fig. 2.10. Eccentric orbit shadowing.**

The optimal perigee can be found simply by comparing the eccentricity vector written in terms of the orbit elements with the meteor radiant vector (recall that the meteor radiant vector is defined as extending from the Earth's center to the radiant point and not the other way around):

$$\hat{L} = [\cos\delta\cos\alpha,\ \cos\delta\sin\alpha,\ \sin\delta]^T \tag{2.17}$$

and

$$\hat{e} = \begin{bmatrix} \cos\Omega\cos\omega - \sin\Omega\sin\omega\cos i \\ \sin\Omega\cos\omega + \cos\Omega\sin\omega\cos i \\ \sin\omega\sin i \end{bmatrix} \tag{2.18}$$

The easiest method of solution is to directly equate the *z*-components of the two vectors such that the desired argument of perigee can be found:

$$\sin\omega\sin i = \sin\delta \tag{2.19}$$

which admits two solutions

$$\omega = \sin^{-1}\left(\frac{\sin\delta}{\sin i}\right) \quad \text{and} \quad \omega = 180 - \sin^{-1}\left(\frac{\sin\delta}{\sin i}\right) \tag{2.20}$$

Figure 2.11 shows the optimal perigee solution, including the equatorial regions for the Leonids and the Draconids. Recall that the optimal node did not have a true solution in the equatorial region; instead, a suboptimal minimum of closest approach existed. The equatorial perigee shown here corresponds to that minimum because the node determines whether shadowing exists, and the perigee simply yields the degree of shadowing for elliptical orbits. Therefore, while not ideal, aligning the perigee point as closely as possible to the approaching storm in this case yields the best solution possible.

The sense of the angles is important to note. For the near-circular case, it did not matter which way the node was pointing when the solution was valid. To generate the solution when flipping the inclination, just switch the ascending and descending nodes. Simply having the plane of the orbit coincident with the

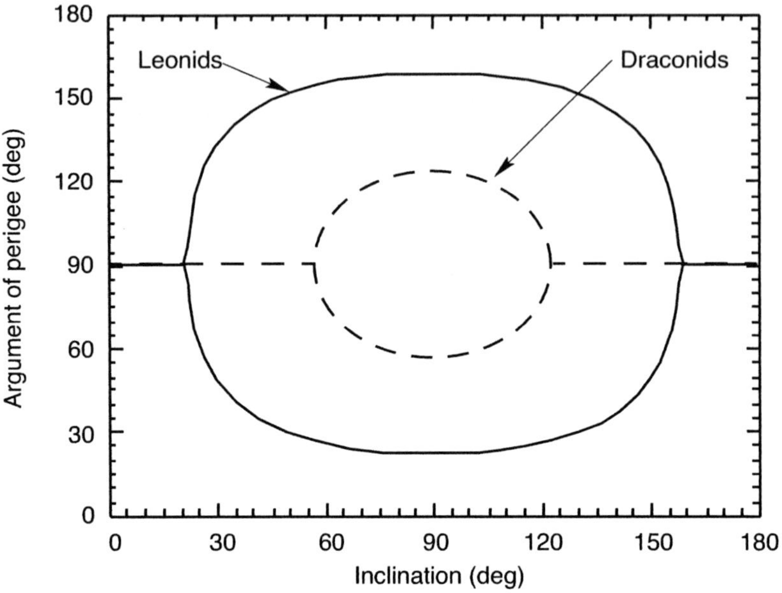


**Fig. 2.11. Argument of perigee for maximal shadowing protection.**

approach vector was enough to generate the optimal solution. When eccentricity is added to the problem, however, the senses of the angle become more significant. The situation is made worse if the perigee is placed into a location different from where it should be and the apogee ends up being pointed toward the shower, resulting in the satellite spending more of its time exposed to the incoming meteors.

For the perigee and the node to be pointed at the storm, the argument needs to be less than 90 (bottom portion of Fig. 2.11). If the node is away from the storm, the argument must be greater than 90, but less than 180 (top portion of Fig. 2.11). No perigee solutions exist where the argument of perigee is greater than 180 deg. Such a case would denote the apogee being pointed at the storm, which is exactly the case to be avoided because the satellite would then spend the vast majority of its time exposed to the meteors. Molniya operators ($\omega$ = 270 deg) need to be aware that major changes are required to the perigee location of those orbits in order to take full advantage of Earth shadowing for the Leonids or for any shower with a positive declination; for negative declinations, the opposite of this argument will hold.

To determine the relative effectiveness of the Earth shadow on eccentric orbits, it is necessary to solve for the time of flight in the safe area of Fig. 2.10. This is more complicated than in the near-circular orbit case, in which the eccentricity could be ignored. To begin, consider the equation of the radius vector of the orbit written in the coordinate system that resides in the plane of the orbit with the $P$ unit vector pointed toward the perigee, $W$ along the angular momentum vector, and $Q$ completes the right-hand system (perifocal coordinate system, Fig. 2.12, $W$ out of the page).

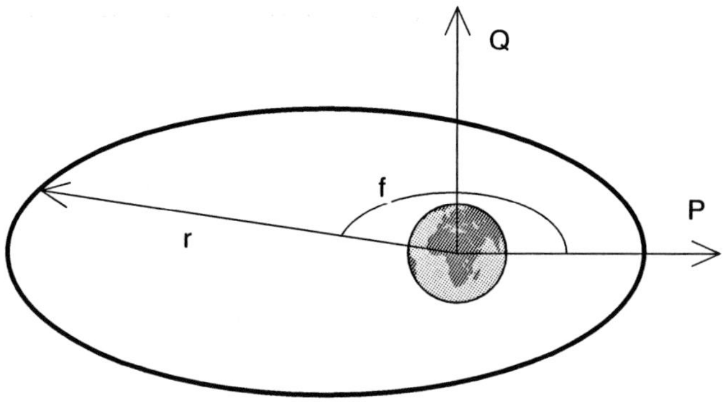


**Fig. 2.12. Perifocal coordinate system.**

The position vector in this system is:

$$\bar{r} = r \cos f\hat{P} + r \sin f\hat{Q} \tag{2.21}$$

where $f$ is the true anomaly.

It is convenient to examine elliptic orbits in terms of the eccentric anomaly $E$ defined as in Fig. 2.13, where the inner ellipse is the orbit, and the circumscribed circle is constructed with the same semimajor axis as the ellipse, centered not on the focal point but on the ellipse center. The angle $E$ is measured from perigee to that point on the circumscribed circle that is met by the perpendicular drawn from the line of apsides through the position vector.

The usual conic equation can be rewritten in terms of the eccentric anomaly as:

$$r = \frac{a(1-e^2)}{1+e \cos f} = a(1-e \cos E) \tag{2.22}$$

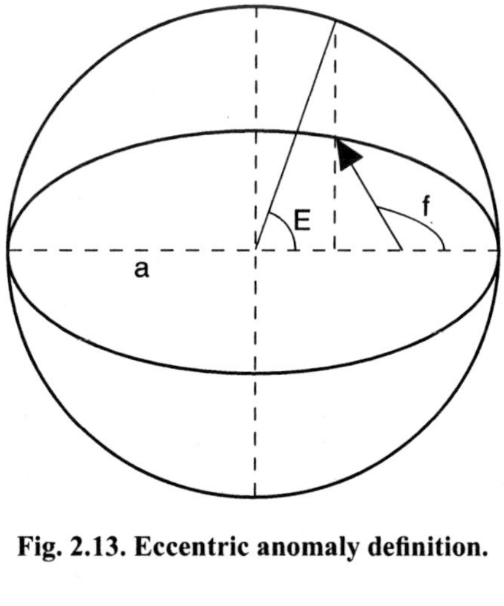


**Fig. 2.13. Eccentric anomaly definition.**

The true anomaly can likewise be rewritten as

$$\cos f = \frac{\cos E - e}{1 - e\ \cos E} \qquad \sin f = \frac{\sqrt{1-e^2}\ \sin E}{1 - e\ \cos E} \tag{2.23}$$

For the problem being examined here, the $Q$ component of Eq. (2.21) is equal to the radius of the Earth's shadow, $R_e + h_a$. Equating the terms gives rise to the expression relating the eccentric anomaly to the shadow size for varying semi-major axis and eccentricity:

$$\sin E = \frac{R_e + h_a}{a\sqrt{1-e^2}} \tag{2.24}$$

This gives two solutions to the eccentric anomaly:

$$E_1 = \sin^{-1}\left(\frac{R_e + h_a}{a\sqrt{1-e^2}}\right) \qquad E_2 = 180 - \sin^{-1}\left(\frac{R_e + h_a}{a\sqrt{1-e^2}}\right) \tag{2.25}$$

However, the eccentric anomaly is defined from the center of the ellipse measured from the perigee and as such, care must be taken in order to use the correct eccentric anomaly. For the case being examined, the larger value of the eccentric anomaly must be used to make sure the shadow is pointing in the right direction toward apogee and not toward perigee (Fig. 2.14).

It is a simple matter, then, of solving for the corresponding mean anomaly $M$ from Kepler's Equation:

$$M_2 = E_2 - e\ \sin E_2 = n\Delta t \tag{2.26}$$

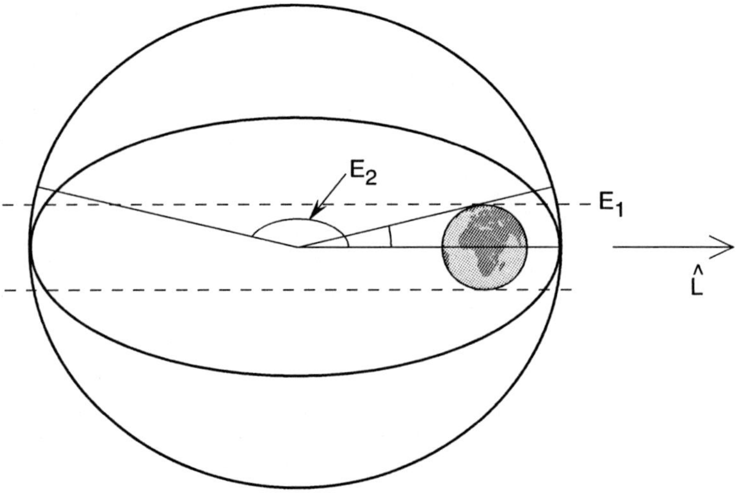


**Fig. 2.14. Correct eccentric anomaly for apogee shadowing.**

Note that the angle of interest is the mean anomaly measured from apogee, not perigee, as it is usually computed. This gives the final adjusted mean anomaly, which is equivalent to the zeta angle $\zeta^*$ as defined during the circular orbit examination (Fig. 2.6), as (where $n$ is the mean motion and $\Delta t_s$ is the time spent in the shadow):

$$\zeta^* = \pi - M_2 = \sin^{-1}\left(\frac{R_e + h_a}{a\sqrt{1-e^2}}\right) + e\left(\frac{R_e + h_a}{a\sqrt{1-e^2}}\right) = \frac{n\,\Delta t_s}{2} \tag{2.27}$$

The percentage of time spent in the shadow is then found in the same manner by comparing the value of $\zeta^*$ to the period of the entire orbit ($\zeta^*$ is measured here in radians to be consistent with Kepler's Equation):

$$\%\text{ protection} = \frac{100}{2\pi}2\zeta^* = \frac{100}{\pi}\left[\sin^{-1}\left(\frac{R_e + h_a}{a\sqrt{1-e^2}}\right) + e\left(\frac{R_e + h_a}{a\sqrt{1-e^2}}\right)\right] \tag{2.28}$$

Equation (2.28) is plotted in Figs. 2.15 and 2.16 as a function of the semimajor axis for various values of the eccentricity in terms of both the percentage of the orbit that is covered and the total time of shadowing. As was the case for the near-circular orbits, the difference between the two concepts is significant. Consider again the 2-hour storm but this time with the highly eccentric Molniya orbits ($e \sim 0.7$, $a \sim 26{,}610$ km). If the timing of the storm is not known, a Molniya orbit can, on average, expect about a 20% reduction in the chances of a meteor impact if the perigee is oriented properly (Fig. 2.15). On the other hand, if the timing of the storm is well determined, the Molniya satellite can be moved in-plane so that the

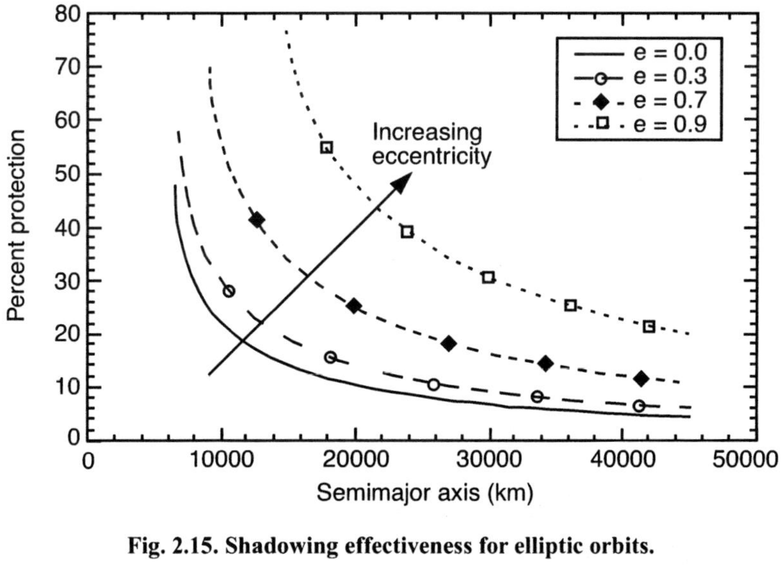


**Fig. 2.15. Shadowing effectiveness for elliptic orbits.**

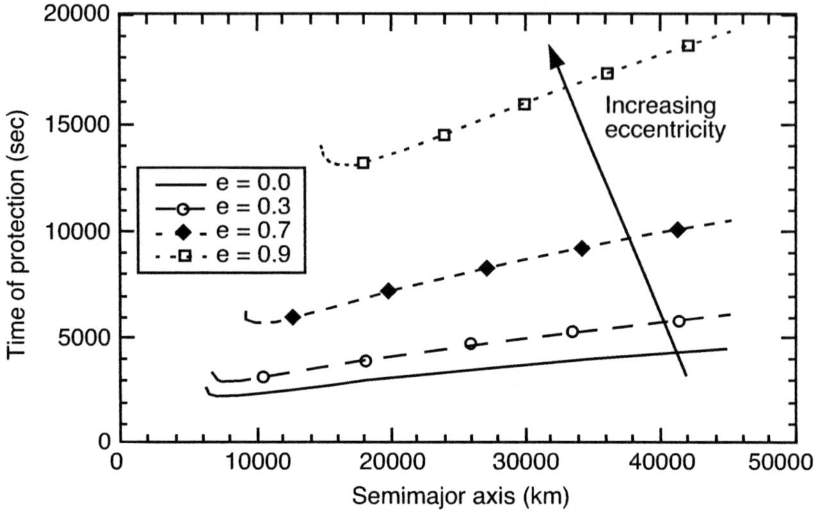


**Fig. 2.16. Total time of protection for elliptic orbits**

protection is maximized to approximately the 8000-sec level (Fig. 2.16). Complete protection in the Earth's shadow is thus possible if the storms are not of excessively long duration and their starting time is known. Even if the start time of the storm is an uncertain variable, eccentric orbits have the potential of providing greater protection than near-circular orbits of the same semimajor axis if they are aligned properly.

1. Optimal node: $$\sin(\Omega_{opt} - \alpha) = -\frac{\tan\delta}{\tan i}$$

2. Optimal perigee: $$\sin\omega_{opt} = \frac{\sin\delta}{\sin i}$$

3. Portion of the orbit that gets shadowed:

$$2\zeta^* = 2\sin^{-1}\left(\frac{R_e + h_a}{a\sqrt{1-e^2}}\right) + 2e\left(\frac{R_e + h_a}{a\sqrt{1-e^2}}\right)$$

4. Time spent in the shadow: $$\Delta t_{shadow} = 2\zeta^*\sqrt{\frac{a^3}{\mu}}$$

**Algorithm 2.3. Optimal eccentric orbit shadowing equations.**

### 2.1.4 Partial Shadowing for Eccentric Orbits

The partial elliptical solution for the node remains the same as it was derived for the near-circular case (Eq. [2.11], Fig. 2.8) modified to account for the change in the semimajor axis. Because the satellite is no longer residing in a constant radius orbit but is assumed to be at apogee in an elliptical orbit, the range of allowable nodes must be modified to account for that altitude. In this case the range of nodes becomes:

$$\sin(\Omega - \alpha) = -\frac{\tan\delta}{\tan i} \pm \frac{R_e + h_a}{a(1+e)\cos\delta\sin i} \qquad (2.29)$$

The near-circular suboptimal analysis relied only upon the unitized angular momentum vector, which does not change with varying eccentricity, and so the only adjustment necessary for elliptical orbits is to account for the greater distance at apogee height. However, as the node moves away from the optimal (i.e., when the plane of the orbit no longer coincides with the approach vector), the optimal perigee location will change as well. In the optimal elliptic case, the $z$-component of the perigee vector was compared directly with the $z$-component of the radiant vector to yield the optimal perigee solution (Eqs. [2.17] to [2.19]). When the orbit is not coincident with the meteor stream, the correct comparison is through the $z$-component of the eccentricity vector and the $z$-component of that vector that comes closest to the approaching storm in the antiradiant direction (i.e., into the shadow). This vector $c$ can be found by crossing the unitized angular momentum into the antiradiant vector, the product of which is crossed back into the unit angular momentum:

$$\hat{c} = (\hat{h} \times -\hat{L}) \times \hat{h} \qquad (2.30)$$

The vector $c$ thus lies in the orbit plane pointing in the antiradiant (i.e., into the shadow) direction and makes the smallest angle possible with the antiradiant vector. Carrying out the mathematical operation gives the closest approach vector as:

$$\hat{c} = -\begin{bmatrix} \cos i(\cos\alpha\cos\delta\cos i - \sin\Omega\sin\delta\sin i) + \cos(\Omega-\alpha)\cos\Omega\cos\delta\sin^2 i \\ \cos i(\sin\alpha\cos\delta\cos i + \cos\Omega\sin\delta\sin i) + \cos(\Omega-\alpha)\sin\Omega\cos\delta\sin^2 i \\ \sin i(\sin\delta\sin i - \sin(\Omega-\alpha)\cos\delta\cos i) \end{bmatrix} \qquad (2.31)$$

Comparing the $z$-components from Eqs. (2.18) and (2.31) (recalling that with this orientation of $c$, the negative eccentricity vector is relevant in order to be consistent with pointing at apogee) gives the suboptimal perigee control law (note that substituting the optimal node will recover the optimal perigee of Eq. [2.19]):

$$\sin\omega = \sin\delta\sin i - \sin(\Omega-\alpha)\cos\delta\cos i \qquad (2.32)$$

Substituting for the available range of the node (Eq. [2.29]) gives the optimal perigee range:

$$\sin\omega = \frac{\sin\delta}{\sin i} \mp \frac{R_e + h_a}{a(1+e)}\frac{\cos i}{\sin i} \qquad (2.33)$$

It must be emphasized that this is the optimal perigee corresponding to a suboptimal node. Most values of the argument of perigee will still yield some semblance of shadowing when the node solution is within range, but for every node angle that results in at least part of the orbit being shadowed, a corresponding optimal perigee will exist. This is given by Eq. (2.33) and Fig. 2.17 for the same LEO and GEO cases examined earlier, with an added eccentricity of 0.1. (Note that the perigee solution is multidimensional in that the perigee is a function of the inclination, semimajor axis, and eccentricity. For this reason a single eccentricity case is shown. Other solutions must be generated for specific values of the orbital elements.)

For the argument of perigee, behavior similar to the node holds. Namely, the GEO satellite has a range of the argument of perigee that is quite small that corresponds to the small range of the node from Fig. 2.8. The small range of possible nodes equates to the small range of optimal perigees for those nodes. LEO satellites, on the other hand, had a very small region of exclusivity for nodal behavior; virtually all nodes provided at least some shadowing. Similar to that is the corresponding behavior for LEO perigees. The wide range of available nodes has a similar wide range of optimal arguments of perigee. Note again that because the perigee lies in the plane of the orbit, most values of the argument of perigee give a shadowed orbit if the node is properly aligned. Equation (2.33) and Fig. 2.17 simply yield the best perigee for the corresponding node for apogee shadowing.

To examine the benefits of partial coverage of elliptic orbits, the arc of the orbit crossing the shadow must be determined. This was a fairly simple operation for partially shadowed near-circular orbits (Eqs. [2.6] and [2.15]) but becomes more complex for the elliptical case. The additional difficulty lies in the

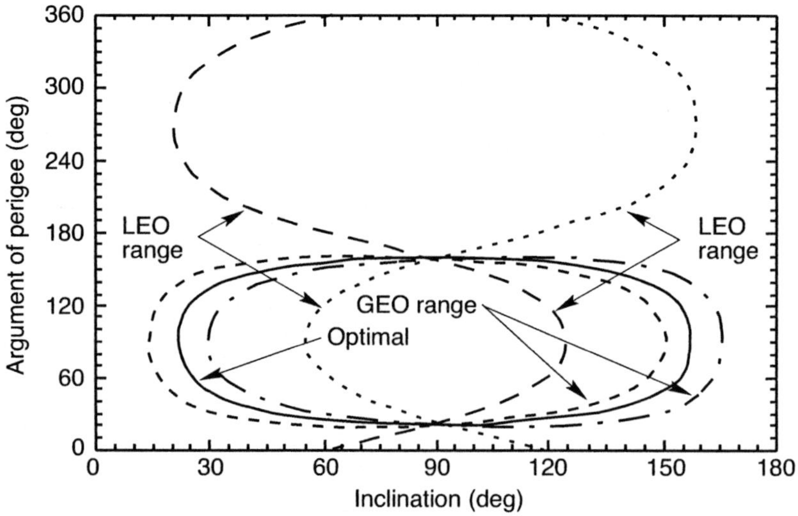


**Fig. 2.17. Leonid optimal perigee for nonoptimal node (e = 0.1)**

fact that the orbit is no longer circular and as such the spherical geometry of Fig. 2.9 (reproduced as Fig. 2.18 for reader's convenience) no longer applies directly to the orbit. But while the orbit itself no longer describes an arc across the surface of a sphere, the spherical relations still hold. The zeta angles for elliptical orbits do not describe true angles but fictitious ones generated to relate the mean motion to the entry and exit points of the shadow. Because the mean motion is by definition constant along a circle of constant radius equal to the semimajor axis of the ellipse, the spherical relations remain valid when applied to the zeta angles. The optimal analysis for elliptical orbits gave rise to Eq. (2.27), which described the time in umbra for an elliptical orbit when the optimal node is paired with the optimal perigee.

But when the node is not optimal, Eq. (2.27) will no longer apply. The orbit plane no longer contains the meteor approach vector. Instead, two options exist: one in which the suboptimal node is paired with the optimal argument of perigee (i.e., the orbit plane is not coincident with the incoming meteor vector, but the perigee is adjusted to be in the best orientation possible), and the second in which neither the node nor the perigee is optimal and both elements are allowed to vary. In this situation, the node range limits still apply, or no shadowing at all would be feasible. The first case to study is the pairing of optimal perigee with the suboptimal node.

In this situation, the node is allowed to move away from the optimal condition, but the perigee is constrained to follow the appropriate control law (Eq. [2.32]). This is the simpler case to examine in that apogee still defines the point of closest approach to the meteor approach vector, and so the problem remains symmetric about the line of apsides. The spherical relations of Fig. 2.18 and Eq. (2.14) still hold if the optimal zeta angle used corresponds to that as derived for elliptical orbits (Eq. [2.27]) and the usual zeta angle continues to define the angle of separation between the orbit plane and the approach vector. By substituting Eq. (2.27) into Eq. (2.14), it can be shown through algebraic manipulation that the new zeta angle becomes (Eq. [2.34]):

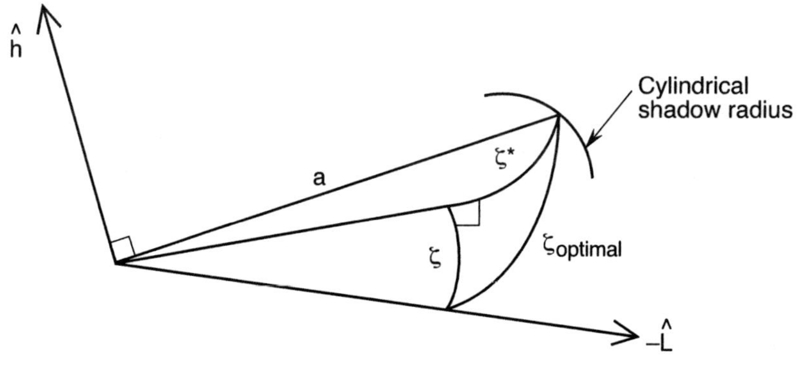


**Fig. 2.18. Orbital arc for eccentric nonoptimal node with optimal perigee.**

$$\zeta^* = \sin^{-1}\left\{\left[\frac{(\gamma\cos e\gamma + \sqrt{1-\gamma^2}\sin e\gamma)^2 - (\hat{L}\cdot\hat{h})^2}{1-(\hat{L}\cdot\hat{h})^2}\right]^{1/2}\right\} \tag{2.34}$$

where

$$\gamma = \frac{R_e + h_a}{a\sqrt{1-e^2}}$$

This reduces to the circular solution when the eccentricity goes to zero and to the optimal elliptic solution when the dot product goes to zero (i.e., when the orbit plane aligns with the meteor vector).

The second, more general solution involves allowing both the node and perigee to drift from the optimal. Hence, the solution becomes transcendental in the eccentric anomaly as per Kepler's Equation (Eq. [2.26]). The first step is to compute the in-plane vector of closest approach as determined in a unitized sense by the vector $c$. Equation (2.29), denoting the range of allowable nodes, presupposed that the apogee was the point of closest approach; this expression would thus give the smallest possible range of nodes because the apogee is farther out than any other point in the orbit. However, if the perigee point is allowed to roam wild and free across the range of space, then the apogee point is no longer applicable when computing the range of useful nodes, and the radial distance at closest approach must be found distinctly.

To compute the radial distance, simply dot product the close approach vector (Eq. [2.31]) into the eccentricity vector (Eq. [2.18]) to generate the cosine of the true anomaly with the sine computed through the dot product of $c$ with the eccentricity vector rotated through 90 deg:

$$\begin{aligned}\cos f_c &= (\hat{c}\cdot\hat{e}) = \sin\omega\cos\delta\cos i\sin(\Omega-\alpha)\\ &\quad -\cos\omega\cos\delta\cos(\Omega-\alpha) - \sin\omega\sin\delta\sin i\\ \sin f_c &= (\hat{c}\cdot\hat{e}_\perp) = \cos\omega\cos\delta\cos i\sin(\Omega-\alpha)\\ &\quad +\sin\omega\cos\delta\cos(\Omega-\alpha) - \cos\omega\sin\delta\sin i\end{aligned} \tag{2.35}$$

where the rotated eccentricity vector is found as:

$$\hat{e}_\perp = \hat{h}\times\hat{e} = \begin{bmatrix}-\cos\Omega\sin\omega - \sin\Omega\cos\omega\cos i\\ -\sin\Omega\sin\omega + \cos\Omega\cos\omega\cos i\\ \cos\omega\sin i\end{bmatrix} \tag{2.36}$$

The radial distance at closest approach can then be found from the conic equation:

$$r_c = \frac{a(1-e^2)}{1+e\cos f_c} \tag{2.37}$$

Substituting into Eq. (2.29) yields the range of available nodes for an arbitrary, free-ranging argument of perigee:

$$\sin(\Omega - \alpha) = -\frac{\tan\delta}{\tan i} \pm \frac{[1 + e\cos f_c](R_e + h_a)}{a(1 - e^2)\cos\delta\sin i} \tag{2.38}$$

The perigee control law of Eq. (2.32) is still valid regardless of whether the node is in an optimal location. As long as the node is maintained within its allowable range, the satellite's orbit will be shadowed at least part of the time irrespective of the argument of perigee. The optimal shading in such a situation can be achieved through following the perigee control law.

To determine the amount of shading that a given orbit has with an arbitrary perigee (i.e., one that does not follow the control law) and a nonoptimal node, the eccentric anomalies of the points where the orbit enters and leaves the shadow must be found in order to solve Kepler's Equation for the time of flight.

By rotating the position vectors for the entrance and exit points through the closest approach angle, the subsequent $y$-components can be directly compared with the radius of the Earth's shadow (Fig. 2.19):

$$-\sin f_c(\cos E_1 - e) + \sqrt{1 - e^2}\cos f_c \sin E_1 = -\frac{R_e + h_a}{a}$$

and (2.39)

$$-\sin f_c(\cos E_2 - e) + \sqrt{1 - e^2}\cos f_c \sin E_2 = +\frac{R_e + h_a}{a}$$

For small eccentricities these equations reduce to Eq. (2.6), which describes the in-plane optimal near-circular solution. For the case where the closest approach is at apogee (the optimal perigee solution, $f_c = 180$), the solution becomes equivalent

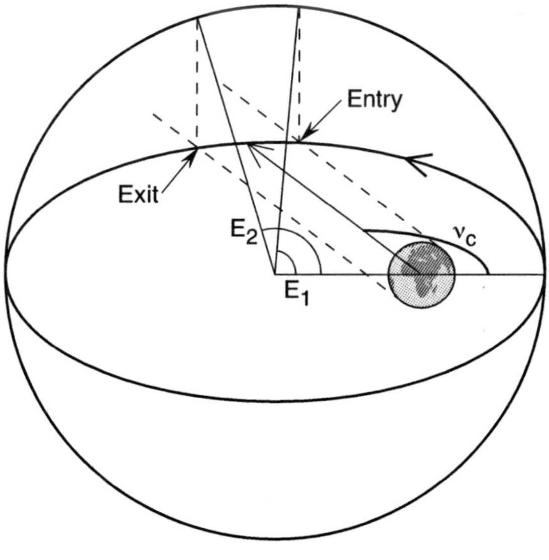


**Fig. 2.19. Nonoptimal node with free-ranging perigee orbit geometry.**

to Eq. (2.23). These expressions need to be solved iteratively to get the eccentric anomalies that directly lead to the zeta angle through Kepler's Equation:

$$2\zeta_{optimal} = (E_2 - e\ \sin E_2) - (E_1 - e\ \sin E_1) = n\Delta t_s \tag{2.40}$$

It must be noted that this is the optimal zeta angle and does not take into account the angular separation of the orbit plane from the meteor approach vector (Fig. 2.19). Instead, it describes the zeta angle that lies in the orbit plane that contains the meteor approach vector for the case where the perigee does not follow its control law (i.e., is arbitrary). The factor of 2 denotes that the angular separation derived from the entrance and exit points is twice the zeta angle as already defined.

The optimal zeta angle again must be solved for iteratively. If the closest approach angle is known, Eq. (2.39) is iterated in the eccentric anomaly, which leads to the solution of Eq. (2.40). This is shown for a representative orbit (a = 10,500 km, e = 0.3) in Fig. 2.20, which shows the time-of-flight in the shadow as a function of the in-plane angle of closest approach. The earlier belief that the apogee point defines the region of maximum shading is confirmed as true.

To account for the angular separation when the orbit plane is not coincident, consider again the optimal zeta angle (Eq. 2.40) and the original zeta angle (Eq. 2.3) giving the amount of angular separation of the orbit plane from the meteor vector. Once the optimal zeta angle is solved for iteratively, the spherical relation of Eq. (2.14) then yields the equivalent angular arc spent in the umbra for elliptical orbits that are not coaligned with the meteor vector:

$$\zeta^* = \sin^{-1}\left\{\left[\frac{\sin^2\zeta_{optimal} - (\hat{L}\cdot\hat{h})^2}{1-(\hat{L}\cdot\hat{h})^2}\right]^{1/2}\right\} \tag{2.41}$$

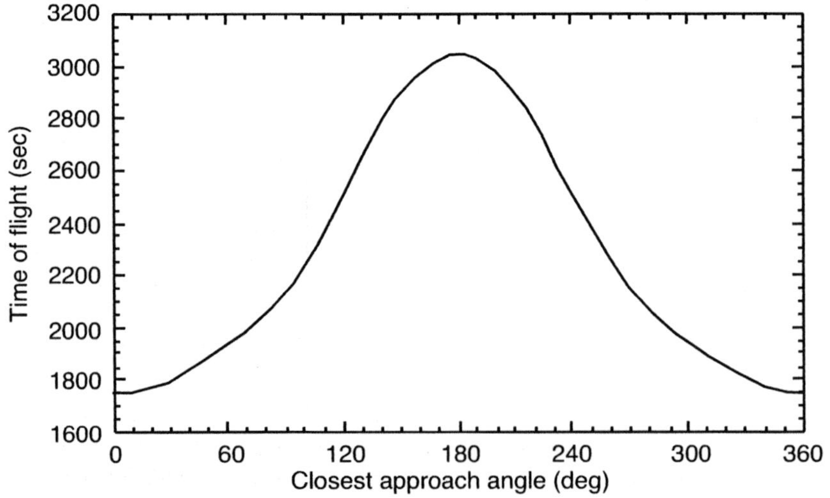


**Fig. 2.20. Time-of-flight in shadow for an eccentric orbit (*a* = 10,500 km; *e* = 0.3).**

1. Optimal node: $\sin(\Omega_{\text{opt}} - \alpha) = -\dfrac{\tan\delta}{\tan i}$

2. Range of nodes: $\sin(\Omega - \alpha) - \sin(\Omega_{\text{opt}} - \alpha) \le \left|\dfrac{R_e + h_a}{a\ (1 + e)\cos\delta\sin i}\right|$

3. Optimal perigee: $\sin\omega_{\text{opt}} = \dfrac{\sin\delta}{\sin i}$

4. Portion of the orbit that gets shadowed when optimal perigee is followed:

$$2\zeta^* = 2\sin^{-1}\left\{\left[\frac{(\gamma\cos e\gamma + \sqrt{1-\gamma^2}\sin e\gamma)^2 - (\hat{L}\cdot\hat{h})^2}{1-(\hat{L}\cdot\hat{h})^2}\right]^{1/2}\right\}$$

where: $\gamma = \dfrac{R_e + h_a}{a\sqrt{1-e^2}}$

5. Portion of the orbit that gets shadowed if optimal perigee is not followed:

$$2\zeta^* = 2\sin^{-1}\left\{\left[\frac{\sin^2\zeta_{\text{optimal}} - (\hat{L}\cdot\hat{h})^2}{1-(\hat{L}\cdot\hat{h})^2}\right]^{1/2}\right\}$$

6. Time spent in the shadow: $\Delta t_{shadow} = 2\zeta^*\sqrt{\dfrac{a^3}{\mu}}$

**Algorithm 2.4. Suboptimal eccentric orbit shadowing equations.**

### 2.1.5 Near-Equatorial Orbits

For near-equatorial, near-circular orbits, there was a limit to the semimajor axis that an orbit could possess and still receive a portion of shading (Eq. [2.12]). This was because the optimal node location was unreachable for near-equatorial orbits (Fig. 2.5). When the orbit has a shallower inclination than the declination of the incoming meteors, the orbit is too shallow for the orbit plane to ever contain the approach vector. The optimal node location of Eq. (2.4) was derived by setting the dot product of the unitized angular momentum vector and the meteor approach vector equal to zero (Eq. [2.3]). For near-equatorial orbits with shadable apogees, the object is to minimize the dot product because the orbit plane can never coalign with the meteor vector. Equation (2.3) can be rewritten for this case, where the $\varepsilon$ is the smallest value achievable:

$$|\hat{h}\cdot\hat{L}| = |\cos\delta\sin i\sin(\Omega - \alpha) + \sin\delta\cos i| = \varepsilon \tag{2.42}$$

The usual range for the values of the declination are between –90 and 90 deg; while for the inclination the typical convention is for the values to be between 0 and 180 deg. Therefore both the cosine of the declination and the sine of the inclination will always be positive. The minimum of Eq. (2.42) then will occur when the sine of the node and right ascension difference is a maximum of one and acts in the opposite sense to the second term. The node control law for near-equatorial orbits is thus:

$$\sin(\Omega - \alpha) = \frac{-\sin\delta\cos i}{|\sin\delta\cos i|} \tag{2.43}$$

with a corresponding range of valid node angles of:

$$\sin(\Omega - \alpha) = \frac{-\sin\delta\cos i}{|\sin\delta\cos i|} \pm \frac{R_e + h_a}{a\ (1+e)\cos\delta\sin i} \tag{2.44}$$

The near-equatorial optimal node, in the case where apogee is shadable, is simply a matter of aligning the node perpendicular to the right ascension of the meteor stream radiant. For prograde orbits, it is necessary to have the node to be following the right ascension by 90 deg, while for retrograde orbits the node should be 90 deg ahead of the right ascension of the meteors. In this manner, the flattest orbit possible with respect to the storm is created (i.e., with maximum shading). This is valid for the Leonids, which approach from above the equator. For meteors coming from below (negative declination), the reverse is true and the prograde node should be 90 deg ahead of the right ascension, while the retrograde ascending node is 90 deg behind.

The best perigee location can be derived in a similar manner. In the optimal case, the eccentricity vector was aligned with the meteor approach vector, but if the orbit plane is not capable of being aligned as in the case of near-equatorial orbits, then the dot product of the eccentricity and meteor vectors will not be equal to one but instead needs to be maximized:

$$(\hat{e} \cdot \hat{L}) = 1 - \varepsilon \tag{2.45}$$

Evaluating the dot product from Eqs. (2.17) and (2.18) gives:

$$\cos(\Omega - \alpha)\cos\omega\cos\delta + \sin\omega\sin\delta\sin i - \sin(\Omega - \alpha)\sin\omega\cos\delta\cos i = 1 - \varepsilon \tag{2.46}$$

Substituting the nodal solution of Eq. (2.43) yields

$$\sin\omega\sin\delta\left(\frac{\cos\delta\cos^2 i}{|\sin\delta\cos i|} + \sin i\right) = 1 - \varepsilon \tag{2.47}$$

Given the conventions of the inclination and declination, the expression in the parentheses will always be positive, which leaves the above relation a maximum when the sine of the argument of perigee is a maximum in the same sense as the declination. The near-equatorial optimal perigee becomes

$$\sin\omega = \frac{\sin\delta}{|\sin\delta|} \tag{2.48}$$

The size of the shadowed arc will be the same for near-equatorial orbits as it was for the suboptimal eccentric orbit case given by Eq. (2.34). This case described the shadowed arc of an elliptic orbit that was not coplanar with the meteor approach vector but was symmetric about the line of apsides. The analysis in computing that relation is equally valid in the near-equatorial case because it, too, is out-of-plane (by definition) and is symmetrical about the optimal perigee. For convenience, the shadowed arc of Eq. (2.34) is given here again as Eq. (2.49).

$$\zeta^* = \sin^{-1}\left\{\left[\frac{(\gamma\cos e\gamma + \sqrt{1-\gamma^2}\sin e\gamma)^2 - (\hat{L}\cdot\hat{h})^2}{1-(\hat{L}\cdot\hat{h})^2}\right]^{1/2}\right\} \tag{2.49}$$

These equations are applicable for eccentric orbits; for the near-circular case simply let the eccentricity go to zero.

1. Optimal node: $\sin(\Omega_{opt} - \alpha) = -\frac{-\sin\delta\cos i}{|\sin\delta\cos i|}$

2. Range of nodes: $\sin(\Omega - \alpha) - \sin(\Omega_{opt} - \alpha) \le \left|\frac{R_e + h_a}{a\,(1+e)\cos\delta\sin i}\right|$

3. Optimal perigee for suboptimal node: $\sin\omega_{opt} = \frac{\sin\delta}{|\sin\delta|}$

4. Portion of the orbit that gets shadowed:

$$2\zeta^* = 2\sin^{-1}\left\{\left[\frac{(\gamma\cos e\gamma + \sqrt{1-\gamma^2}\sin e\gamma)^2 - (\hat{L}\cdot\hat{h})^2}{1-(\hat{L}\cdot\hat{h})^2}\right]^{1/2}\right\}$$

where: $\gamma = \frac{R_e + h_a}{a\sqrt{1-e^2}}$

5. Time spent in the shadow: $\Delta t_{shadow} = 2\zeta^*\sqrt{\frac{a^3}{\mu}}$

**Algorithm 2.5. Equatorial orbit shadowing equations.**

## 2.2 Lunar Shadowing

The Earth is not the only body for which shadowing is possible. The Moon also exists close to the Earth and may be able to shadow portions of a satellite's orbit if the lunar geometry is correct. For example, if the Moon is between the Earth and the approaching meteors, it may be able to shield the vehicle; if it is on the other side away from the approaching storm, it is not useful for shadowing purposes (Fig. 2.21).

The Moon is moving quite rapidly with respect to the satellite in comparison with the Earth, which the satellite sees as stationary. Therefore, the date and time of meteor storm passage is crucial in determining whether the Moon can act as a

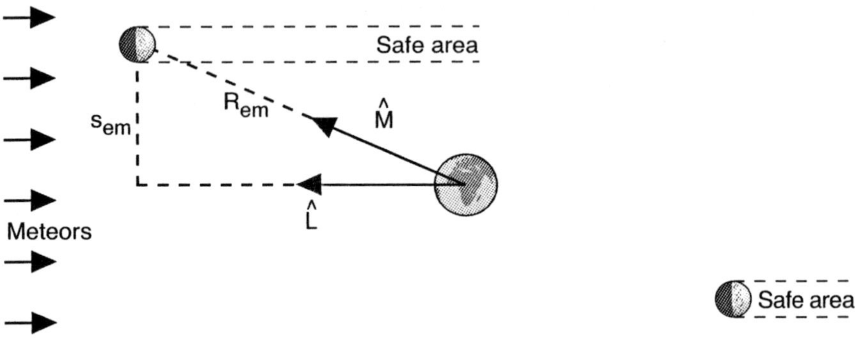


**Fig. 2.21. Meteoric eclipse cast by the Moon.**

beneficial obstacle to meteor passage. Table 2.1 shows the angular lunar positions as given by the NASA Jet Propulsion Laboratory (JPL) ephemerides for the next several years during the anticipated times of peak Leonid activity.

It is a simple matter, then, to take the lunar position vector and compare it with the meteor approach vector to obtain the separation distance that denotes the perpendicular distance between the projected lunar shadow and the Earth's center (Fig. 2.21). Note that for purposes of simplifying the computation, the finite size of the Moon is ignored, and a point-mass is assumed that casts a cylindrical shadow equal in radius to the Moon's diameter. If the solution comes close to any specified orbit, the increased complexity added when considering a finite Moon can be examined. The separation distance is computed as

$$s_{\mathrm{em}} = R_{\mathrm{em}}\sqrt{1-(\hat{L}\cdot\hat{M})^2} \tag{2.50}$$

where $R_{em}$ is the distance from the Earth to the Moon, $M$ is the unit vector from Earth to Moon, and $L$ is the meteor approach vector. For lunar shadowing to be a viable alternative, the separation distance must be less than the orbit apogee. If that occurs, the orbit angles can be altered to take advantage of the Moon's presence.

The results from Eq. (2.50) are shown in Fig. 2.22, which gives the separation distance $s_{em}$ during the times immediately before and after the anticipated Leonid period of peak activity on November 17 for the critical years. For any orbit that

**Table 2.1. Lunar Position During Anticipated Leonid Storms**

| Date | Right Ascension | Declination |
|---|---|---|
| 18 h Nov. 17, 1997 | 6 h 24 min 03.2576 sec | +17 d 55 min 05.849 sec |
| 18 h Nov. 17, 1998 | 14 h 35 min 19.6432 sec | -10 d 47 min 20.753 sec |
| 6 h Nov. 18, 1999 | 23 h 17 min 46.1512 sec | -08 d 33 min 33.456 sec |
| 18 h Nov. 18, 2000 | 22 h 05 min 34.0364 sec | +18 d 25 min 55.979 sec |

has an instantaneous distance from the Earth of values larger than these curves, lunar shadowing is possible. However, through examination it is apparent that the Moon does not come close enough during the times of the Leonids to be a viable candidate for the vast majority of Earth orbiters. If we take the minimum distance as shown to be 44,000 km during the year 2000 passage, we can see that this just barely touches upon the orbits of geosynchronous satellites, and only for a few hours during one day at that. This justifies the earlier assumption of treating the Moon like a point-mass for the computations. Even adding in the Moon's radius of 1738 km, Fig. 2.22 shows that the Moon is not a viable candidate for beneficial shadowing.

## 2.3 Applications to LEO Satellites

The general assumption in the space community has been that satellites residing in orbits close to the Earth will receive protection from the Earth and its atmosphere amounting to close to 50% of the orbit. However, for shadowing to take place, the satellite has to be in such an orbit that it spends at least part of its time in the shadow. If the orbit plane is perpendicular to the oncoming stream, it would not matter how low the altitude is; the satellite would never be in the Earth's shadow. It would not be correct to assume that simply because the orbit is at a low altitude that it will receive much protection; it is quite possible that depending upon the orbital geometry, it will not receive any protection at all. The purpose of this section, then, is to examine LEO satellites and determine where the optimal protection and regions of exclusion (i.e., areas with no protection) reside in order for satellite operators to know where they want their satellites to be or not to be.

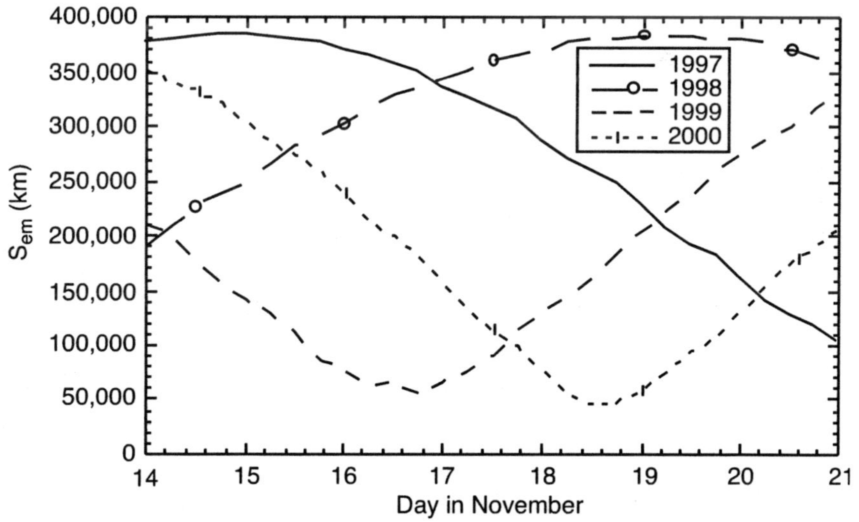


**Fig. 2.22. Separation distance between projected lunar shadow and center of the Earth.**

The amount of protection that a satellite will receive will be dependent upon the relative geometry between the orbit and the vector from which the meteors are approaching:

$$\%=\left(\frac{100}{\pi}\right)\sin^{-1}\left[\frac{\gamma^2-(\hat{L}\cdot\hat{h})^2}{1-(\hat{L}\cdot\hat{h})^2}\right]^{1/2} \tag{2.51}$$

Figures 2.23 to 2.26 show the percent protection for a range of values of LEO semimajor axis as functions of the inclination and node at the time of the meteor storm. To get an idea of the approximate level of protection that a given satellite will receive from the storms, simply cross-reference the plot applicable to the closest semimajor axis with the inclination of the satellite and the node at the time of the storm.

All the figures are characterized by a wide, relatively flat region around the optimal case, punctuated by two separate areas of steep change that are centered around a region of zero protection shadowing effect. It is thus apparent that even with very low Earth orbiting satellites, a region exists wherein no protection at all is afforded to the vehicle (as shown by black areas on the plots). Physically, the center points of the regions of exclusion correspond to those orbits having a plane perpendicular to the approach vector of the incoming meteors (see Fig. 2.3). As the orbit grows in size, it can sustain a greater and greater variation from the perpendicular case and still receive no protection. Thus, as the semimajor axis grows, so does the region of exclusion. Care must then be taken by satellite operators to be aware that these regions exist, and if Earth shadowing is desired, the vehicles are maneuvered to stay away from them.

The limits of the regions of exclusion can be found from simple geometry. For shadowing to occur, the angular distance from the approach vector multiplied by the radius must be less than the cylindrical extent of the umbra (Fig. 2.27):

$$a\left|\hat{L}\cdot\hat{h}\right| \le R_e + h_a \tag{2.52}$$

When the angular extent becomes too large, the orbit loses the possibility of shadowing. Conversely, the region of exclusion is defined simply by the reverse of Eq. (2.52):

$$a\left|\hat{L}\cdot\hat{h}\right| > (R_e + h_a) \tag{2.53}$$

In terms of the orbit elements, the dot product is found from Eq. (2.3); therefore, Eq. (2.53) admits two solutions:

$$\cos\delta\sin i\sin(\Omega-\alpha)+\sin\delta\cos i>\gamma$$

and

$$-\cos\delta\sin i\sin(\Omega-\alpha)-\sin\delta\cos i>\gamma \tag{2.54}$$

where $\gamma$ was earlier defined as the ratio of the Earth's effective meteor shadow to the semimajor axis of the orbit.

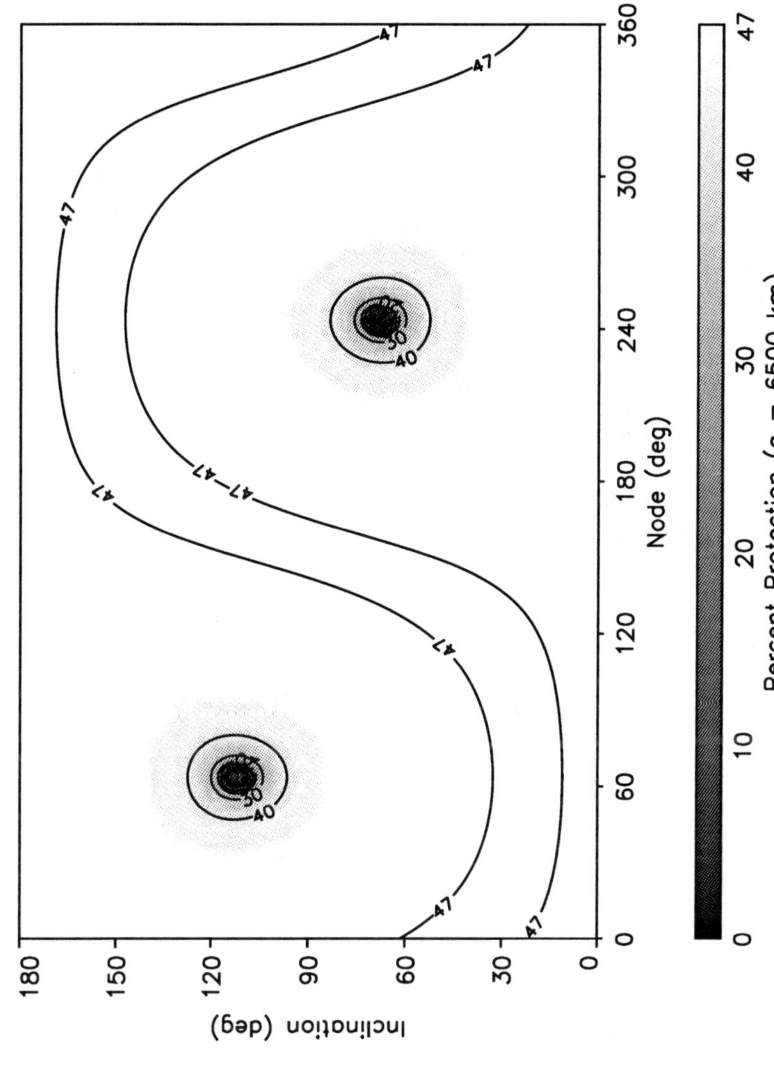


**Fig. 2.23. Earth shadowing protection for 6500 km LEO orbit.**

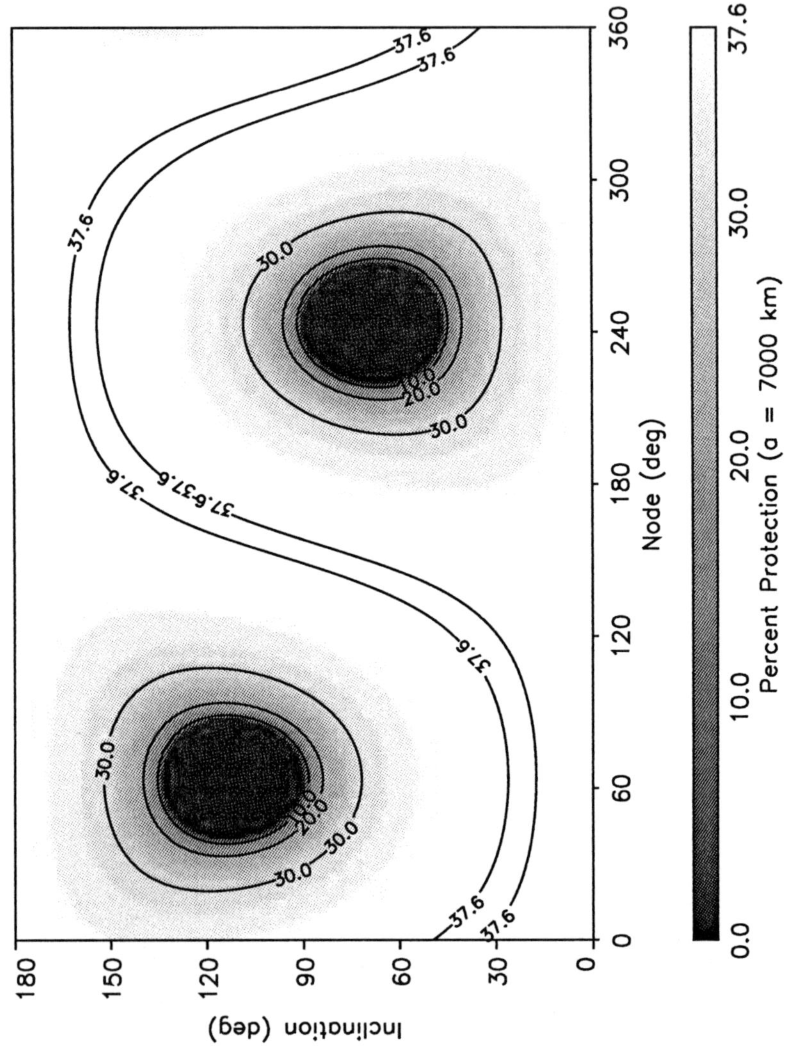


**Fig. 2.24. Earth shadowing protection for 7000 km LEO orbit**

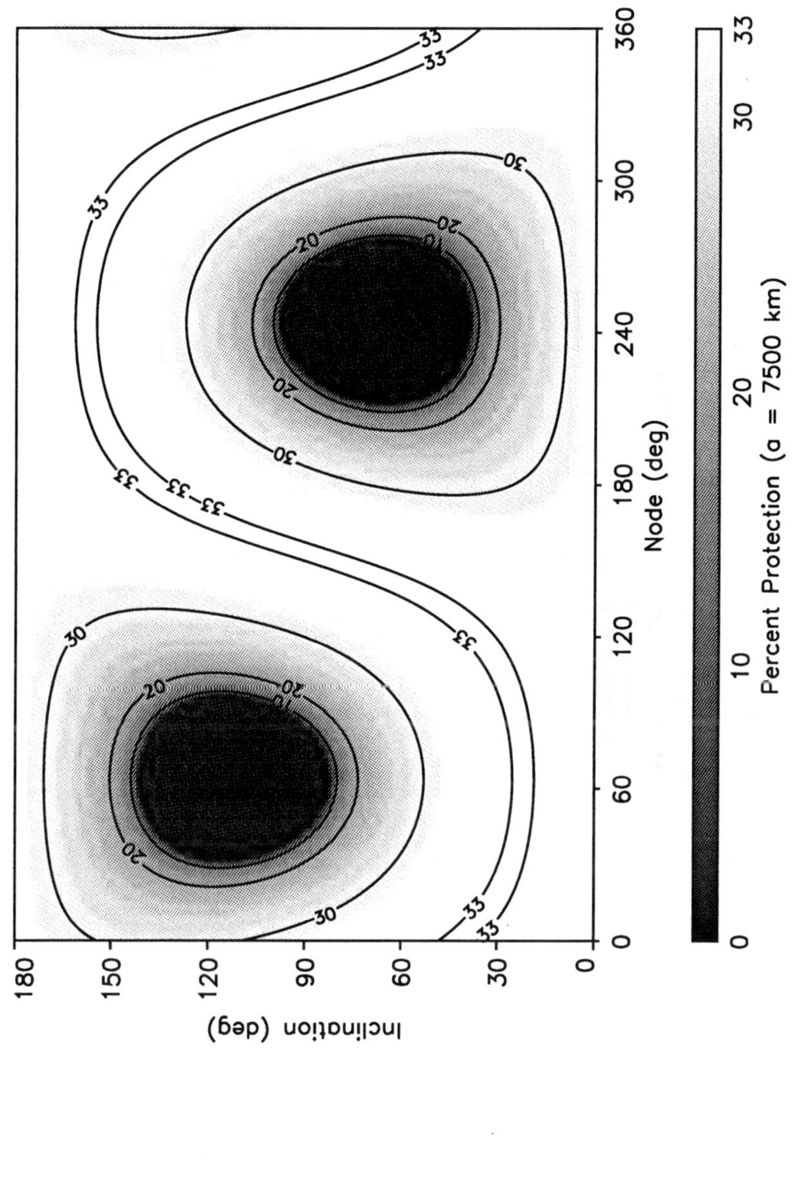


**Fig. 2.25. Earth shadowing protection for 7500 km LEO orbit.**

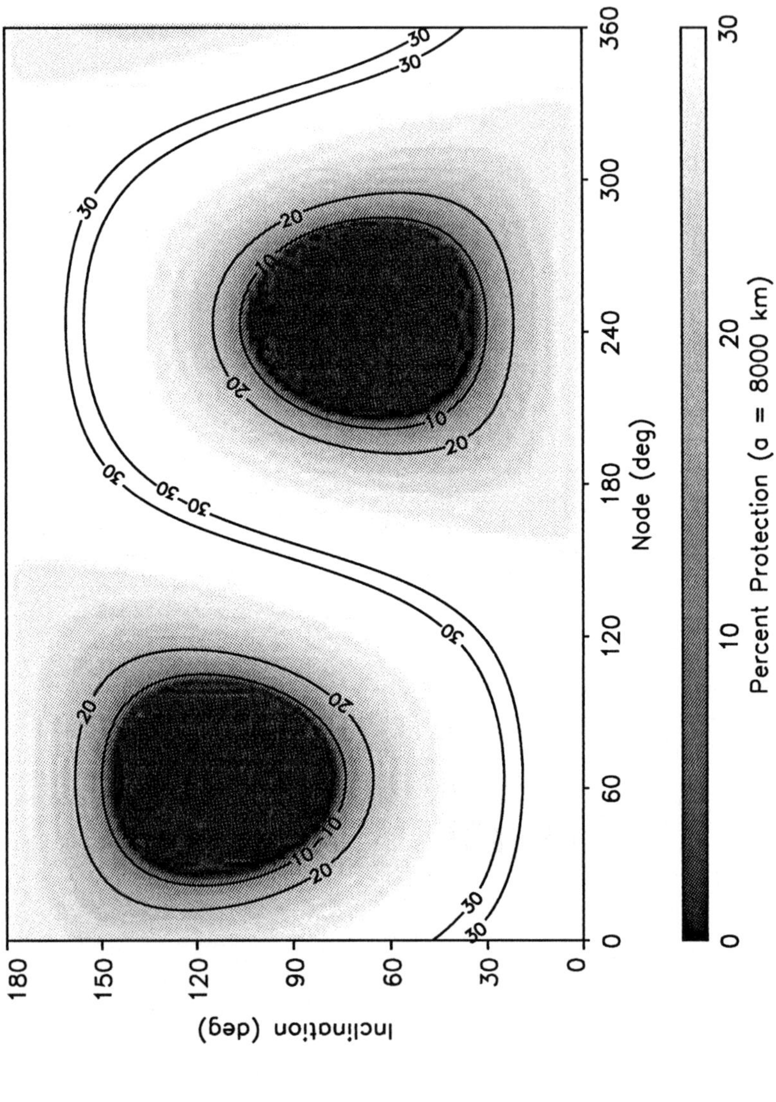


**Fig. 2.26. Earth shadowing protection for 8000 km LEO orbit.**

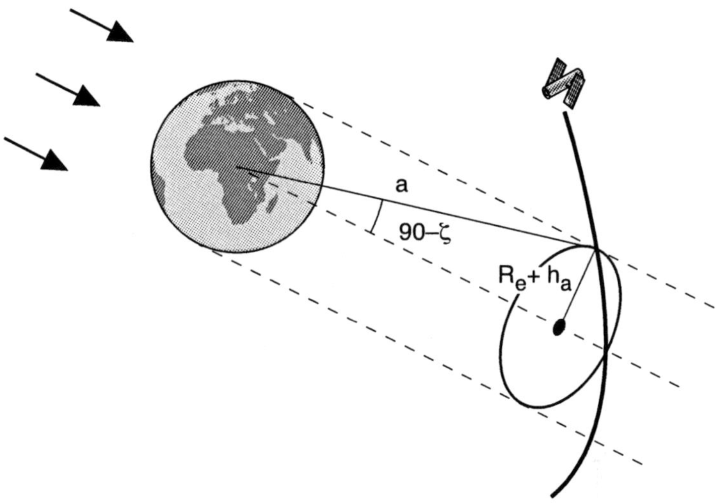


**Fig. 2.27. Angular extent of Earth shadowing.**

Equation (2.54) gives the interrelation between the semimajor axis, inclination, and node at the time of the storm, determining whether any shadowing occurs at all. Because the node is constantly precessing as a result of the Earth's oblateness, that orbital element varies with time, with the worst case occurring when the line of nodes is perpendicular to the equatorial projection of the meteor's approach vector (right ascension of the radiant point). Because it is desired to find those orbits that may include a region of exclusion, the worst case for the node will be assumed. The inclination and semimajor axis, then, are the crucial solve-for parameters in determining whether the orbit has the chance of missing shadowing benefits. Equation (2.54) reduces to:

$$\begin{aligned} &\text{for } \sin(\Omega - \alpha) = 1: && \sin(i + \delta) > \gamma \quad \text{or} \quad \sin(i + \delta) > -\gamma \\ &\text{for } \sin(\Omega - \alpha) = -1: && \sin(i - \delta) > \gamma \quad \text{or} \quad \sin(i - \delta) > -\gamma \end{aligned} \tag{2.55}$$

Solving for the inclination gives:

$$\begin{aligned} &\text{for } \sin(\Omega - \alpha) = 1: && \sin^{-1}(\gamma) - \delta < i < 180 + \sin^{-1}(-\gamma) - \delta \\ &\text{for } \sin(\Omega - \alpha) = -1: && \sin^{-1}(\gamma) + \delta < i < 180 + \sin^{-1}(-\gamma) + \delta \end{aligned} \tag{2.56}$$

Equation (2.55) corresponds to the maximum limits of the regions of exclusion in Figs. 2.23 to 2.26, assuming the worst case location for the node (recall that the right ascension of the Leonid radiant point is 153.45 deg). Figure 2.28 shows the limits of Eq. (2.55) for the Leonids in graphical form. It is a simple matter, then, to cross-reference a given semimajor axis with the corresponding inclination in order to determine whether or not the satellite operator will ever face a condition of nonshadowing.

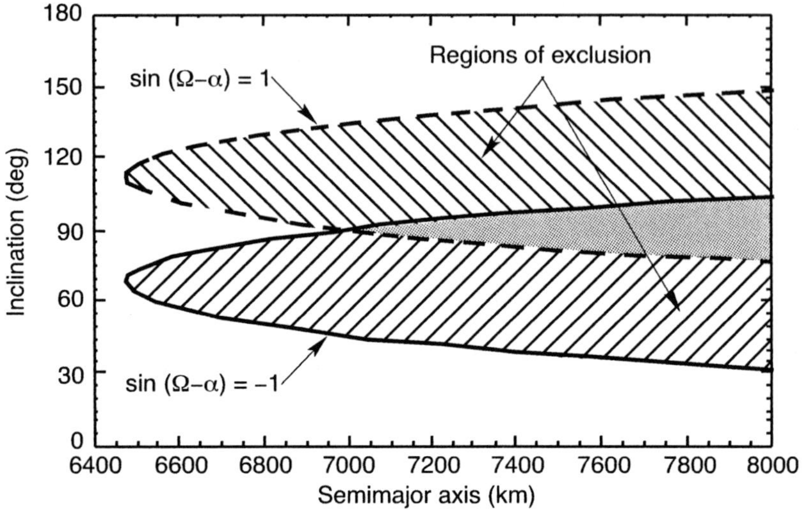


**Fig. 2.28. Leonid regions of exclusion.**

With the limits of the regions of exclusion established, certain general orbit types are of specific interest. Many LEO satellites reside in polar (or near-polar) and Sun-synchronous orbits. Equatorial and frozen orbits are also fairly common. Each of these types is briefly examined next, along with the only verified case of a LEO satellite being hit by a meteor (the 1991 Solar A Perseid event).

### 2.3.1 Polar Orbits

For purely polar satellites (those satellites assumed to have an inclination of exactly 90 deg), Eq. (2.53) can be solved in terms of the semimajor axis (because the declination ranges between –90 and +90 deg, the cosine of the declination can be divided without changing the sign of the inequality):

$$|\sin(\Omega - \alpha)| > \frac{\gamma}{\cos\delta} \tag{2.57}$$

For the worst-case node location, the right ascension of the ascending node is perpendicular to the right ascension of the radiant point. The semimajor axis can then be directly solved for, giving:

$$a > \frac{R_e + h_a}{\cos\delta} \tag{2.58}$$

Thus, for polar satellites at a semimajor axis greater than 6986 km (assuming an effective atmosphere of 100 km), there is a risk of the satellite not receiving any Earth shadowing protection at all (also seen in Fig. 2.28) for the Leonids. Care then needs to be taken in this situation to avoid those right ascensions of ascending nodes that result in no shadowing.

### 2.3.2 Equatorial Orbits

For satellites in equatorial orbits, the inclination goes to zero, the dependency upon the node is eliminated, and Eq. (2.53) can again be solved directly, giving the limiting value of the semimajor axis for those orbits (valid for any node location):

$$a > \frac{R_e + h_a}{|\sin\delta|} \tag{2.59}$$

The smallest Leonid value of the semimajor axis satisfying this constraint is 17,293 km for an effective atmosphere assumed 100 km above the physical Earth. For LEO satellites in near-equatorial orbits, there exists no risk of not having some semblance of Earth shadowing. However, it does imply that GEO equatorial orbits are not going to receive any Earth shadowing at all. Indeed, if the reverse problem is solved for GEO satellites, it can be seen that those spacecraft will not receive any naturally occurring shadowing unless the declination of the approaching meteors is less than 8.8 deg.

### 2.3.3 Frozen Orbits

Frozen orbits are defined as those orbits that have a perigee location that is fixed in space. Because Earth oblateness is the main driver in the secular perigee motion, the inclination is fixed in such orbits at 63.43 deg or 116.57 deg in order to drive this term to zero (Wertz and Larson, 1991):

$$\dot{\omega}_{J_2} = \frac{3}{4} n J_2 \left(\frac{R_e}{a}\right)^2 \frac{(4 - 5\sin^2 i)}{(1 - e^2)^2} \tag{2.60}$$

where $n$ is the mean motion, $J_2$ is the Earth oblateness, and $a$ is the Earth's equatorial radius. Equation (2.53) then reduces to (where the plus/minus sign refers to whether the orbit is prograde or retrograde):

$$\left|\sin(\Omega - \alpha) \pm \frac{\tan\delta}{2}\right| > \frac{\sqrt{5}}{2} \frac{\gamma}{\cos\delta} \tag{2.61}$$

Solving Eq. (2.60) for the worst relative nodal angles between the orbit and the radiant point yields:

$$a > \sqrt{5} \frac{R_e + h_a}{|2\cos\delta \pm \sin\delta|} \tag{2.62}$$

which gives two limiting values. The lower value corresponds to the limiting value being sought because that will dictate if a loss of shadowing is possible. For the Leonids, the limiting value of the semimajor axis is 6498 km. From Fig. 2.28 it can be seen that the frozen orbit is close to the worst inclination possible (63.43 deg and 116.57 deg frozen inclination versus 68.13 deg and 111.87 deg Leonid worst-case inclination). Therefore, for frozen orbits of virtually any altitude, a risk exists for shadow loss.

### 2.3.4 Sun-Synchronous

Sun-synchronous satellites are governed by a secular precession in the node caused by the Earth's oblateness of 0.9856 deg/day in order to keep the orbit plane aligned with the Sun's position:

$$\dot{\Omega}_{\text{sun synch}} = -\frac{3}{2}nJ_2\left(\frac{R_e}{a}\right)^2\frac{\cos i}{(1-e^2)^2} = 0.9856 \text{ deg/day} \tag{2.63}$$

For near-circular satellites, this reduces to an expression relating the semimajor axis and the inclination (Wertz and Larson, 1991):

$$0.9856 = -2.06474 \times 10^{14} a^{-7/2} \cos i \tag{2.64}$$

Equation (2.64) can be substituted into Eq. (2.53) to generate the expression for the regions of exclusion for Sun-synchronous satellites. The resulting expression can be solved for numerically, giving 6712 km as the minimum semimajor axis at which it is possible to lose Earth shadowing for the Leonid shower.

However, most Sun-synchronous satellites follow the Sun in order to keep as much of the orbit in sunlight as possible, implying that the orbit plane will be close to perpendicular with respect to the Sun's direction. Given that the relative geometry between the Earth's orbit at the time of the Leonids and the direction from which the meteors are approaching is naturally close to perpendicular, it is apparent that even though the possibility exists for Sun-synchronous orbits to miss shadowing benefits, the typical operational status of such orbits already requires the node to be close to the optimal location. Unless the satellite is following the Sun for reasons other than to maximize solar exposure, Sun-synchronous orbits will most likely not lose shadow.

### 2.3.5 Solar A and the Perseids

While two known satellites have been lost to meteor collision (Olympus I and Solar A), only Solar A had the potential of being protected by the Earth's shadow. Although not much of a conclusion can be drawn from one data point, it is interesting to note that if the level of protection provided to Solar A from the Earth's umbra for the Perseid showers is computed from Eq. (2.51), only about a 15.35% level of protection is achieved (assuming an approximate Perseid radiant point of $\alpha = 46$ deg, $\delta = 58$ deg, Appendix B). Solar A did not get much help from the Earth's shadow and thus spent an inordinate amount of time exposed to the incoming particles.

If the regions of exclusion are examined for the Perseids in a similar manner as were those just completed for the Leonids, Fig. 2.29 is generated. It can be seen that orbits with high inclinations will not completely lose shadowing protection until the satellite is far away from the Earth. Conversely, lower-inclined orbits have a greater risk of umbra loss. Compare this with Fig. 2.28 for the Leonids. In that case, orbits with more extreme inclinations are subject to greater risk of shadow loss because the meteors are approaching from the relatively shallow angle of 22 deg, while the lower orbits are not subject to such a risk.

For the Perseid meteor shower, the limiting values of the semimajor axis for the earlier-examined special cases are 12,225 km for polar orbits, 7634 km for equatorial orbits, 7592 km for frozen orbits, and 8635 km for Sun-synchronous orbits, assuming an effective atmosphere of 100 km. Again, one data point does not provide a good basis for comparison, but Solar A, with a semimajor axis of 7004 km and inclination of 31 deg, corresponds closely to the worst-case scenario of Fig. 2.29.

## 2.4 Applications to Constellations

Constellation protection raises a question potentially separate from individual satellites. Specifically, while placing a satellite's plane in the optimal condition maximizes that plane's protection, it does not take into account the other planes that make up the constellation. The optimum for a constellation may be different from the single-plane case. The question with constellations is not to maximize protection for the individual planes but rather to determine if there are optimal node locations whereby more than one plane can be shadowed. As a result, the constellation's protection becomes larger than the shadow protection afforded to a single plane. Therefore the optimal node previously derived applies as optimal only for single planes; for constellations the optimal may actually be offset from this value.

It is a fairly simple matter to alter the equations already derived. In order to determine whether a single plane is in the shadow, Eq. (2.10) was used; to find whether multiple planes of a constellation are in the shadow, that expression must be altered to include the other planes:

$$-\frac{R_e + h_a}{a} \le \cos\delta \sin i \sin[\Omega - \alpha + (n-1)\Delta\Omega] + \sin\delta \cos i \le \frac{R_e + h_a}{a} \tag{2.65}$$

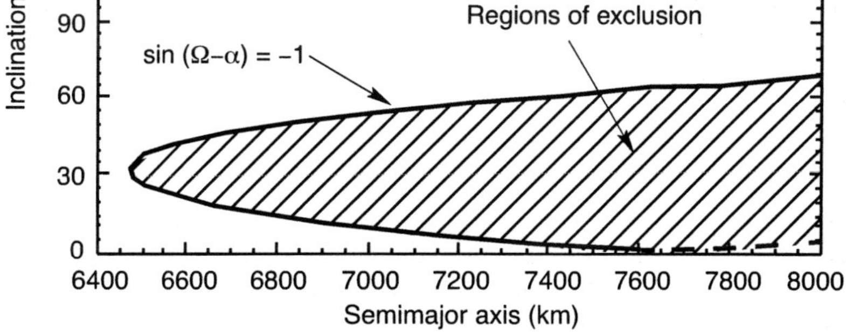


**Fig. 2.29. Perseid regions of exclusion.**

where $n$ is the plane number (i.e., $n$ goes from 1 to 6 for GPS) and $\Delta\Omega$ is the nodal spacing. Each plane can then be examined to determine if it is in the shadow or not. Once a plane is found to be in the shadow, Eq. (2.15) can be used to compute the fraction of the orbital arc that is shadowed for that plane.

Finally, the number of satellites in each of the planes must be considered. For example, if 30% of the orbit is shadowed and the plane has five satellites, there is a probability of 1 that at least one satellite will be in eclipse, with an additional 50% probability that a second satellite will also be shadowed. Assuming the satellites are evenly spaced in the plane (separated by a constant in-plane angular interval $\Delta f$), the number of satellites definitively in the shadow is:

$$SV_{\text{shadow}} = \text{INT}\left(\frac{2\zeta^*}{\Delta f}\right) \tag{2.66}$$

The probability of an additional satellite (or a single satellite if $SV_{\text{shadow}}$ is zero) being in the shadow is:

$$P_{\text{shadow}} = \text{MOD}\left(\frac{2\zeta^*}{\Delta f}\right) \tag{2.67}$$

A statistical measure of the overall protection provided to the constellation by the Earth is then simply the total number of satellites definitively in the shadow, plus the probability that *any* additional satellite is also eclipsed. (Permutations need not be considered because this analysis is interested in quantifying the overall statistical sense of the protection as a function of the node angle.) The relevant statistical parameter becomes:

$$F_{\text{shadow}} = \sum_n SV_{\text{shadow}_n} + 1 - \bigcap_n (1 - P_{\text{shadow}_n}) \tag{2.68}$$

The best solution of Eq. (2.68) over the $\Delta\Omega$ range (i.e., the optimal node for the constellation) can then be compared with the single-plane optimal case in order to understand the relative benefits between optimizing the protection for the single plane and the full constellation. Reiterating, the optimal node location for the single plane may differ from the full constellation solution both in degree of protection and nodal position. The overall process of analyzing constellation protection is summarized in Algorithm 2.6.

Note in examining the constellation, the range of right ascension of ascending nodes need be quantified by ranging the node of the first plane only from 0 to $\Delta\Omega$, rather than through the entire 360 deg. The assumed symmetry in nodal spacing in the constellation allows the entire system to covered in this manner since the only concern is not in which of the planes are shadowed but only in how many planes. However, because every constellation that will be examined has a different $\Delta\Omega$, then for practical purposes of figure generation the entire range of the node from 0 to 360 will be presented in the following analysis.

Seven currently (or partially) aloft constellations are examined as examples for this study. Table 2.2 lists the relevant orbit parameters for each of these constellations. Orbcomm consists of a main group of satellites in a lower-inclined orbit supplemented by an additional plane inclined more toward the pole at 70 deg.

For each node angle (letting the node of the first plane range from 0 to $\Delta\Omega$, symmetry will cover the rest of the constellation):

1. Determine which of the planes are in shadow:

$$-\frac{R_e + h_a}{a} \le \cos\delta \sin i \sin[\Omega - \alpha + (n-1)\Delta\Omega] + \sin\delta \cos i \le \frac{R_e + h_a}{a}$$

2. Determine the fraction of the orbital arc for each shadowed plane:

$$2\zeta^* = 2\sin^{-1}\left\{\left[\frac{\left(\frac{R_e + h_a}{a}\right)^2 - (\hat{L} \cdot \hat{h})^2}{1 - (\hat{L} \cdot \hat{h})^2}\right]^{1/2}\right\}$$

3. Compute the probability of a satellite being in the shadow for each plane:

$$SV_{\text{shadow}} = \text{INT}\left(\frac{2\zeta^*}{\Delta f}\right) \qquad P_{\text{shadow}} = \text{MOD}\left(\frac{2\zeta^*}{\Delta f}\right)$$

4. Combine each planar probability:

$$F_{\text{shadow}} = \sum_n SV_{\text{shadow}_n} + 1 - \bigcap_n (1 - P_{\text{shadow}_n})$$

5. Compare with the optimal single-plane case:

$$\sin(\Omega_{\text{opt}} - \alpha) = -\frac{\tan\delta}{\tan i} \qquad 2\zeta^* = 2\sin^{-1}\left(\frac{R_e + h_a}{a}\right)$$

$$F_{\text{optimal}} = SV_{\text{optimal}} + P_{\text{optimal}} = \left(\frac{2\zeta^*}{\Delta v}\right)$$

**Algorithm 2.6. Constellation and single-plane shadow protection.**

**Table 2.2. Orbital Characteristics of Examined Constellations**

| Constellation | Number of planes | $\Delta\Omega$ (deg) | *SV* | $\Delta f$ (deg) | *a* (km) | *i* (deg) |
|---|---|---|---|---|---|---|
| Orbcomm | 4 | 90 | 8 | 45 | 7200 | 45 |
| OrbcommP | 1 | na | 4 | 90 | 7200 | 70 |
| Iridium | 6 | 60 | 11 | 32.727 | 7150 | 86.4 |
| GlobalStar | 8 | 45 | 6 | 60 | 7780 | 52 |
| Teledesic | 12 | 30 | 24 | 15 | 7750 | 84.7 |
| ICO | 2 | 180 | 5 | 72 | 16735 | 45 |
| GLONASS | 3 | 120 | 8 | 45 | 25520 | 64.8 |
| GPS | 6 | 60 | 4 | 90 | 26610 | 55 |

Because the highly inclined orbit consists of a single plane, a constellation analysis for OrbcommP is not applicable. However, the single-plane analysis for OrbcommP will be carried through the initial computations for comparison purposes. Neither the ICO nor Teledesic constellations is deployed, although Teledesic has launched an experimental satellite and plans more such satellites before the end of the Leonid threat in November 2000. All the others are currently either operational (Iridium, GPS, GLONASS, Orbcomm) or in the deployment stage (GlobalStar). As such, these constellations may feel the full brunt of the Leonid situation and therefore are relevant to this study.

Table 2.3 gives for each of the constellations the optimal node location, the maximum orbital arc protection corresponding to the optimal node (values greater than 360 simply wrap around), and the node angle ranges for which shadowing is possible but not necessarily optimal. (All values are in degrees.)

Table 2.3 is reproduced in graphical form in Figs. 2.30 and 2.31, which represents the statistical parameter $F$ for the single-plane case ($F$ is linearly scalable to the shadowed arc length $\zeta^*$, Algorithm 2.6) and provides a basis for comparison with the full constellation results examined next. The factor $F$ indicates that for the lower constellations, if the single plane is aligned with the optimal condition, then at least one and possibly more satellites are guaranteed to be in the eclipse when the meteor storm occurs (for example, Teledesic would have at least 7 of the 24 satellites in shadow and a 56% chance of an 8th). The higher-altitude MEO constellations (ICO, GLONASS, GPS) have far fewer satellites with corresponding larger spacing; in addition they have a smaller portion of the orbital arc covered and are therefore more susceptible to loss of shadow. An $F$ of zero indicates that the orbit plane receives no shadowing at all. Geometrically, the center of this region corresponds to the situation when the orbit plane is perpendicular to the oncoming meteors. As the altitude grows, the region of exclusion

**Table 2.3. Shadow Characteristics of Examined Constellations for Single-Plane Case**

| Constellation | Optimal Node | Node Range | Maximum Arc ($2\zeta^*$) | $F_{\text{single plane optimal}}$ |
|---|---|---|---|---|
| Orbcomm | 129, 357 | 258–589 | 128.24 | 2.85 |
| OrbcommP | 145, 342 | 272–575 | 128.24 | 1.425 |
| Iridium | 152, 335 | 262–585 | 129.92 | 3.96 |
| GlobalStar | 135, 352 | 278–569 | 112.74 | 1.88 |
| Teledesic | 151, 336 | 84–213<br>274–403 | 113.41 | 7.56 |
| ICO | 129, 357 | 72–164<br>323–415 | 45.55 | 0.63 |
| GLONASS | 143, 344 | 125–159<br>327–362 | 29.40 | 0.65 |
| GPS | 137, 350 | 117–155<br>332–10 | 28.18 | 0.31 |

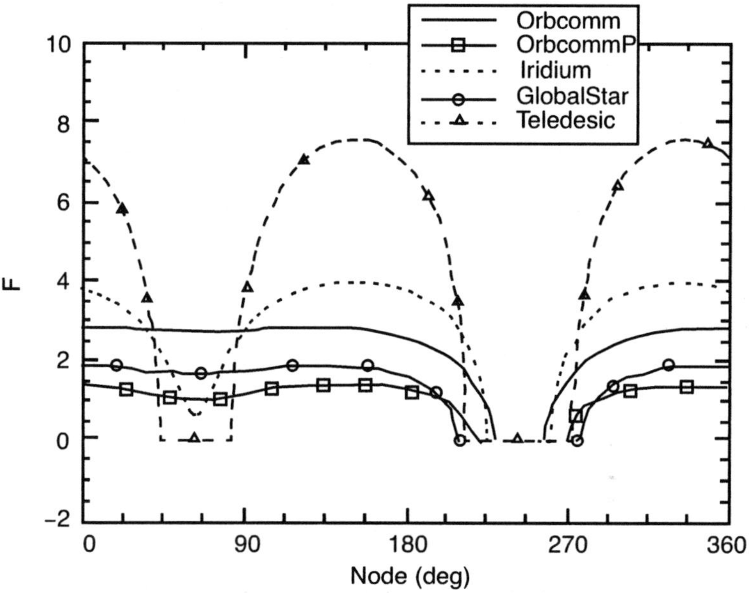


**Fig. 2.30. Shadow parameter *F* for single-plane case for LEO constellations.**

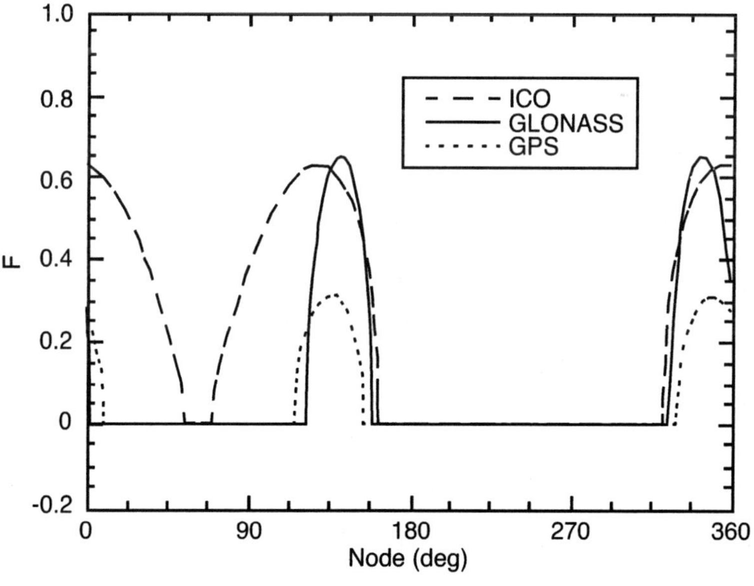


**Fig. 2.31. Shadow parameter *F* for single-plane case for MEO constellations.**

will grow until it dominates the problem at MEO-type altitudes (ICO, GLONASS, GPS) (Fig. 2.31).

For the full constellations, the problem becomes more complex. The parameter *F* as shown in Algorithm 2.6 computes the same for constellations as it does for the single-plane case, namely that the integer portion of *F* represents the number of satellites certain to be in the shadow for any given node, plus a statistical probability that at least one other vehicle is eclipsed (decimal fraction of *F*). This is shown in Figs. 2.32 to 2.34. Note that as expected from symmetry of the constellation, the periodicity of the figures is equivalent to the number of planes in the constellation. For example, Iridium consists of six planes and therefore has a periodicity over a 360-deg node range of 6. Also, it should be noted that (for the LEO constellations) the statistical parameter *F* looks to be an integer. In fact it is not, but it is very close to an integer value. The probability in these cases of at least one additional satellite being in eclipse is very close to one, owing to the large number of satellites and planes in the constellation

Table 2.4 shows the optimal node and *F* parameter for the constellations (Figs. 2.32 to 2.34) compared with the same constellation with one of its planes fixed in the optimal single-plane orientation (Eq. [2.4]). In other words, the constellation is fixed at either the optimal node for the full constellation or at the single-plane optimal node. The *F* parameters are then computed for the full constellation in both cases in order to make an accurate comparison. The *F* parameter for the single-plane case here again is not the same as in Table 2.3, which showed only the individual plane. Here, the single-plane *F* is computed for the full constellation but wherein one of the planes is fixed to the single-plane optimal node value from Table 2.3.

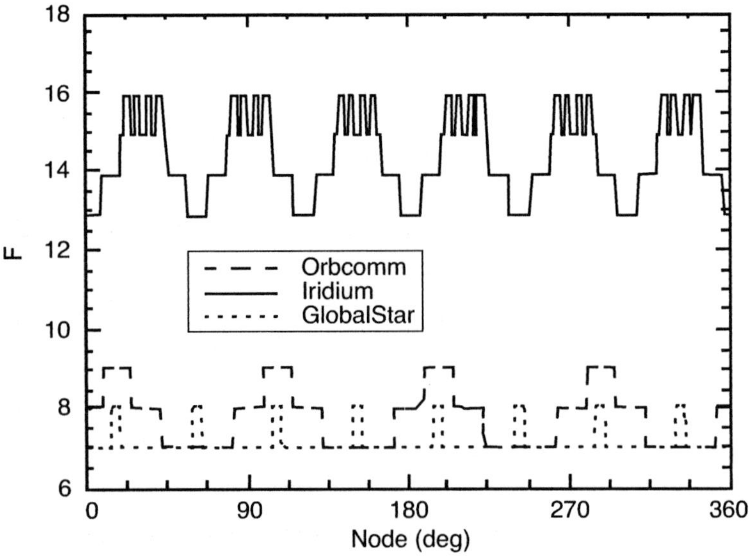


**Fig. 2.32. Shadow parameter *F* for full LEO constellation.**

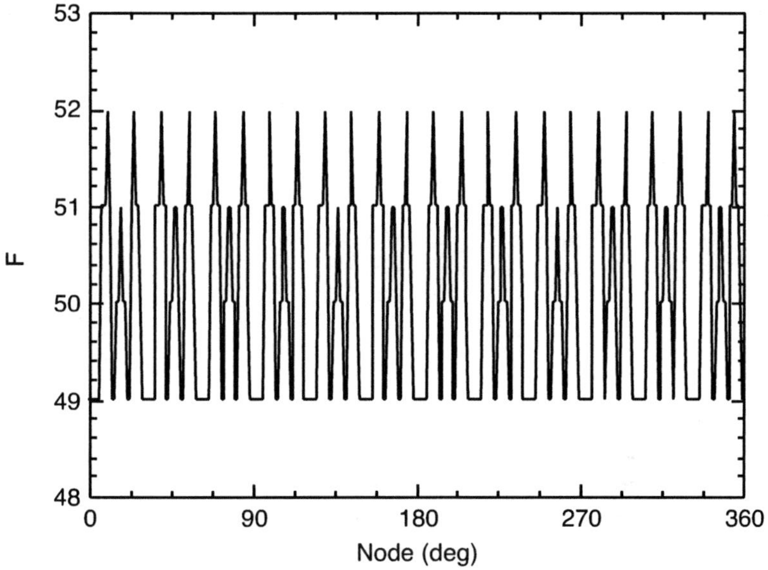


**Fig. 2.33. Shadow parameter *F* for full Teledesic constellation.**

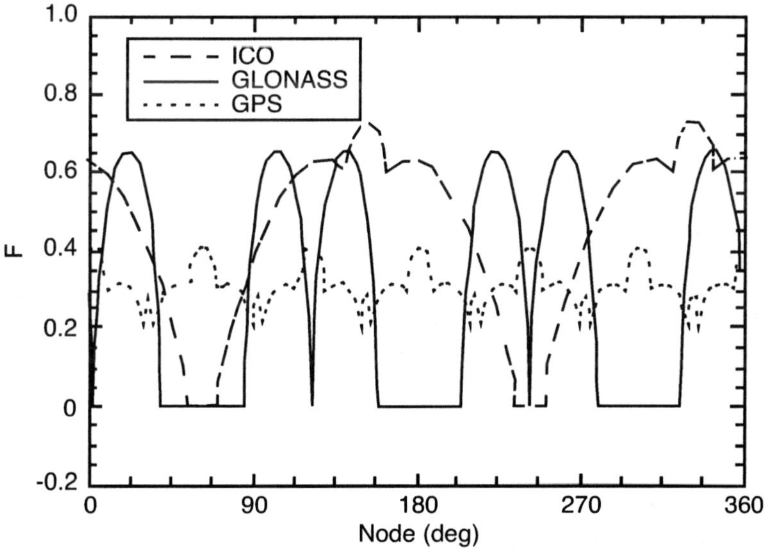


**Fig. 2.34. Shadow parameter *F* for full MEO constellation.**

**Table 2.4. Shadow Characteristics for Full Constellations**

| Constellation | Optimal node for single plane | $F_{sp}$ | Optimal node for constellation | $F_{min}$ | $F_{max}$ |
|---|---|---|---|---|---|
| Orbcomm | 129 | ~8 | 110-115 | ~7 | ~9 |
| Iridium | 152 | ~15 | 143-163 | ~13 | ~16 |
| GlobalStar | 135 | ~7 | 151-155 | ~7 | ~8 |
| Teledesic | 151 | ~49 | 145,160 | ~49 | ~52 |
| ICO | 129 | 0.633 | 152.5 | 0 | 0.724 |
| GLONASS | 143 | 0.653 | 103,143 | 0 | 0.653 |
| GPS | 137 | 0.313 | 122 | 0.212 | 0.408 |

The optimal node constellation values are listed for those maximums closest to the optimal node single-plane case (i.e., the optimal node for Orbcomm is 110 to 115 deg; although other maximums exist, this is the maximum closest to the 129-deg single plane optimal node value). For Orbcomm and GlobalStar, the values are flat and equal to the maximum over the given range. For Iridium, the range is sawtooth in nature, alternating every 2 to 3 deg with an $F$ of 15. Teledesic also shows the sawtooth type of behavior largely because of the higher inclination and the large number of planes. With singular distinctive maximums at 145 and 160 deg, the maximums given for Teledesic are not as sawtooth as Iridium (close to the 151-deg single-plane optimal node, other maximums exist). The minimums for the LEO constellations are all over unity, indicating that regardless of the node angle of the constellation, some satellites will always be shadowed.

The MEO constellations have more distinctive maximums and are smoother in nature than the LEO configurations, but none of the three examined here have $F$ greater than unity. This indicates that regardless of the node angle, the MEO constellations will never definitely have a satellite in shadow. It is useful to note that the constellation maximums for GLONASS are the same as they are for the single-plane case. Both GLONASS planes cannot be in the shadow at the same time, and therefore the optimum for the constellation is the same as that for the single-plane case. ICO has some overlap between planes, producing a larger maximum than the single-plane case at a different nodal location. But a minimum $F$ value of zero means that for some node locations, both ICO and GLONASS have absolutely no chance of being in shadow. GPS always has at least approximately a 20% probability of having one of its vehicles in eclipse, but unlike the LEO constellations, this probability is only a positive chance and never a surety. The GPS constellation maximum is also larger than for the single-plane node location and, like ICO, resides at a different nodal position than the single-plane optimal node.

## 2.5 Summary of Shadowing

This chapter presented many complex orbital mechanics, which were condensed into a series of algorithms (2.1 to 2.6) that should make the equations easier to apply. The relevant equations for meteoric eclipses were derived, but with a simple substitution of the Sun's radiant for the meteor radiant, the equations are equally applicable for solar eclipses as well. The optimal node location represents the orbit plane being in alignment with the incoming meteors; the satellite thus passes directly behind the Earth and obtains maximum shadowing (Eq. [2.4]). Because the orbit normal vector does not depend upon the eccentricity, its value remains the same regardless of the orbit shape. The node control law for equatorial orbits represents an achievable node that will provide the best shading possible for those orbits incapable of being coaligned (Eq. [2.43]). These are the target nodes for a satellite operator to try to achieve for lone satellites. For constellations, the optimal nodes can be different from the single satellite case and are solved numerically. The range of nodes gives the limits that the right ascension of ascending node can assume if any shadowing is to take place and represents the allowable error in trying to achieve the optimal node (Eq. [2.29]).

Altering the perigee is of course good only for those orbits that have noticeable perigees, namely eccentric orbits. The optimal argument of perigee corresponds to that angle that, when coupled with the optimal node, gives the longest duration of Earth shadowing by placing the apogee of the orbit directly behind the Earth with respect to the arriving meteors. The perigee control law presents the optimal solution (Eq. [2.19]). Also derived was a suboptimal solution that applies when the node is not at its optimal location, but is within the range where shadowing does exist (Eq. [2.32]). The perigee control law for near-equatorial orbits simply gives the argument of perigee that comes closest to the desired vector alignment with the approach vector (Eq. [2.48]).

The arc size expresses the level of shadowing present for a given orbit shown here for optimal and suboptimal, and circular (Eqs. [2.6] and [2.15]) and elliptic orbits (Eqs. [2.27) and [2.34]). The zeta angle is the measure of how much of the orbit arc is placed within the shadow of the Earth. For elliptical orbits, this does not represent the actual angular extent of the shadowed arc but instead the equivalent arc when mean motion is assumed (i.e., the time spent in umbra). The elliptical orbit arc solutions are for the cases where both the node and perigee control laws are followed (the optimal) and where the node law is not followed but the perigee law is (suboptimal). For the elliptical orbits where the perigee law is not followed, the free-ranging perigee case is a transcendental equation and not expressible in unique form. The zeta angle is thus found iteratively through Eqs. (2.39) to (2.41). For near-equatorial orbits, the solution is similar to the suboptimal elliptic case just described. But whereas that solution was suboptimal, for near-equatorial orbits that solution for the arc represents the best that can be achieved because planar coalignment is not possible. Therefore, the optimal arc size for near-equatorial orbits is the same as for the suboptimal elliptic case (Eq. [2.49]). If the control laws are not followed for near-equatorial orbits, then the arc size solution must be solved iteratively using Eqs. (2.39) to (2.41).

# 3 Gravitational Focusing of Meteoroid Particles

Getting a gravitational assist from planetary flybys has long been a means of boosting spacecraft into more advantageous orbits. The Galileo mission and its Venus-Earth-Earth-Gravity-Assist (VEEGA) profile is one of the more complex examples of the flyby phenomena. However, all objects, not just spacecraft, passing in the vicinity of a larger body will experience a similar effect; this would apply to such small particles as meteoroids as well. The resulting trajectories have the potential of being significantly altered from their pre-encounter condition as a consequence of the flyby. This chapter examines the flyby phenomena for meteoroid streams and how they change not only the shadow that is cast by the larger body but also the flux threat that an orbiting spacecraft faces.

## 3.1 Flyby Geometry

When particles emitted from a meteor shower encounter the Earth, they burn up in the atmosphere. As just examined in Chapter 2, this situation causes a cylindrical gap to appear in the density of the shower on the opposite side of the planet from which the meteor particles are approaching. If a satellite's orbit is constrained to partially inhabit that particular region of space, the vehicle receives an increased amount of safety from potentially dangerous meteoroid encounters (Fig. 3.1).

The geometric analysis performed in Chapter 2 assumes that the shadow is of a consistent cylindrical shape, with the radius of the cylinder equal to the radius of the Earth, plus any protective atmosphere. The problem is not as simple as this, however. The meteor stream is approaching from interplanetary space, so even though individual particles may have slight orbital deviations from their local brethren, overall the stream is going to be moving constantly from a known direction with respect to the Earth. Once the particles encounter the Earth's sphere of influence, the Earth, being a gravitating body, will cause the particles to change path as they stream by. With the high-approach velocity of

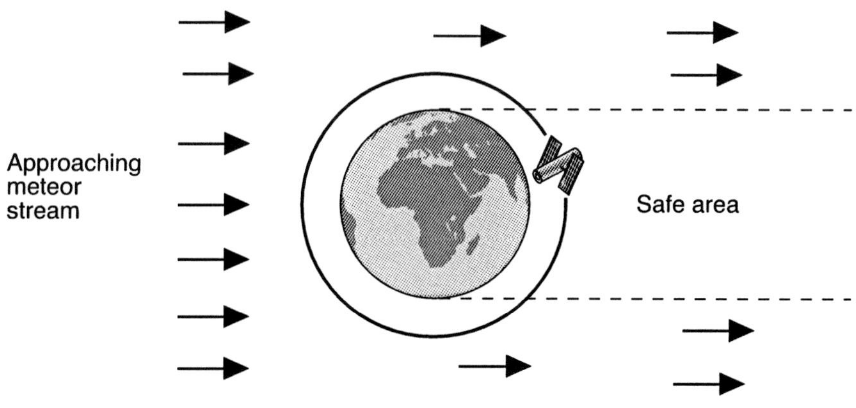


**Fig. 3.1. Meteoric eclipse cast by the Earth.**

the incoming particles, capture into an elliptical orbit is not possible for the meteoroids. Instead, the problem represents a hyperbolic flyby for each traveling particle, the geometry of which is shown in Fig. 3.2. The meteors are here approaching the Earth with a velocity of $V_\infty$ (~70 km/sec for the Leonids). The Earth's mass rotates the particles through the turning angle δ; the meteors then leave with the same velocity but in a new direction.

Since the presence of the Earth causes the meteor stream to alter its local orbit, there will come a point whereby the meteors streaming past the Earth on either side will cross paths. Assuming the Earth is a spherical body, the curvature changes will be symmetric about the approach vector (meteor shower radiant point) in three dimensions. The paths just missing the Earth will meet at a focal point on the opposite side of the Earth, as denoted in Figs. 3.2 and 3.3 by the vector *d*. Closer to the Earth than this is a region where the particles have been intercepted and the remaining particles cannot reach. This is the area of complete shadowed protection from the shower. Outside of this region, however, the flux density will still not be equal to the free-space approaching flux.

The areas denoted by a single asterisk (*) in Fig. 3.3 will see the normal particle density only slightly changed, while that regime indicated by double asterisks (**) represents a crossover region receiving a meteor density that may be orders of magnitude greater than usual. Because the Earth is causing a three-dimensional symmetrical focusing, the resulting particle density along the meteor

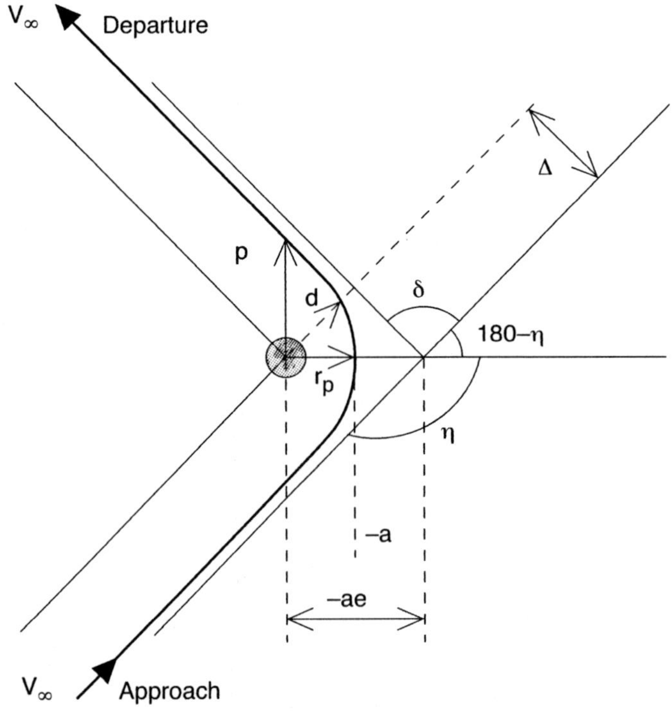


**Fig. 3.2. Hyperbolic orbit geometry.**

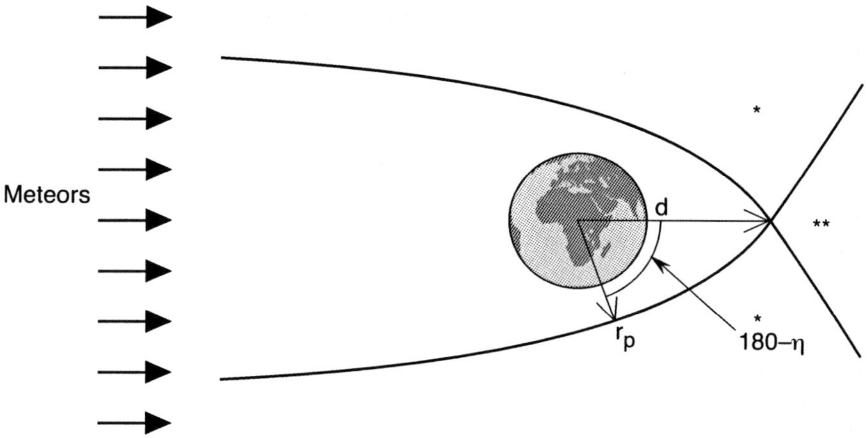


**Fig. 3.3. Dust particle approach and Earth shadowing.**

shower antiradiant direction (the vector $d$) represents the greatest combined density possible in the ** region. Therefore, it is crucial to determine the location of the focal distance and the resulting density along the antiradiant direction.

If Fig. 3.2 is reexamined, $d$ simply represents the continuation of the approach vector, positioned so it goes through the center of the Earth. Therefore the angle that $d$ makes with respect to the perigee vector $r_p$ (point of closest approach) is given by 180–η. The expression relating the eccentricity of the orbit to η is (Bate *et al.*, 1971):

$$\cos(180 - \eta) = +\frac{1}{e} \tag{3.1}$$

Given the general equation for a conic:

$$r = \frac{p}{(1 + e\cos f)} \tag{3.2}$$

where $p$ is the semilatus rectum and $a_\infty$ is the semimajor axis of the flyby orbit:

$$p = a_\infty(1 - e^2) \tag{3.3}$$

then the focal distance $d$ can be solved for as (where the true anomaly $f$ is given by the desired angle 180–η):

$$d = \frac{p}{\left(1 + e \times \frac{1}{e}\right)} = \frac{p}{2} \tag{3.4}$$

Similarly, the eccentricity of the flyby orbit can be expressed as (Ginsberg and Luders, 1997):

$$e = 1 + \frac{r_p V_\infty^2}{\mu} \tag{3.5}$$

The smallest value of the eccentricity (and hence the greatest amount of resulting curvature) can be found by assuming the distance of closest approach is equivalent to the Earth's radius. Because the value of the Leonid approach velocity is fixed ($V_\infty$ = ~70 km/sec), a minimum eccentricity of 79.405 is achieved for the passing particles. This value of the eccentricity will increase linearly as the distance of closest approach increases.

By substituting Eqs. (3.4) and (3.5) into Eq. (3.3), the crossover distance for an arbitrary particle that is not constrained to just barely pass by the Earth's surface becomes

$$d = r_p + \frac{1}{2}\frac{V_\infty^2}{\mu}{r_p}^2 \tag{3.6}$$

where again $r_p$ is the distance of closest approach (perigee of the flyby orbit).

For the Leonid meteors, this results in a crossover distance of 256,000 km; clearly the focusing effect is not an issue for this shower. But the reverse problem can be solved as well. Most Earth-centered satellites will never be higher than geosynchronous altitude (semimajor axis of 42,240 km). Substituting this semimajor axis into Eq. (3.6), an approach velocity of 26.5 km/sec is found. Only meteor showers of this approach velocity or less will be influenced by the focusing to such an extent that satellite operators need to be concerned. Still, the analysis will be presented in a general context with the specific example of the Leonids being treated; the GEO-shower case will also be discussed.

## 3.2 Instantaneous Flux after Earth Encounter

Because the particles' trajectories are changing as a consequence of the flyby, the local values of the instantaneous fluxes will change as well. Qualitatively, there will be five distinct regions of flux values: the free-space pre-Earth encounter flux (Sec. 1.4), the shadowed region (protected area of zero flux), the flux along the antiradiant vector (maximum possible), the post-encounter partially focused flux in the near-Earth environment (* region in Fig. 3.3), and the crossover region, where partially focused particles will cross over the paths of other particles coming from the other side of the Earth that have already crossed the antiradiant vector (** region in Fig. 3.3). Each of these areas is different in the flux behavior. This section will examine the flux along the antiradiant, the flux in the near-Earth environment before crossover occurs, and the flux after crossover but away from the antiradiant.

### 3.2.1 Flux Along the Antiradiant

The most intense region of the particle flux created by the focusing effect will be along the antiradiant direction. But how bad will it be? Consider two particles separated by a small amount $\delta r$ (Fig. 3.4). The distance of closest approach between the flyby orbits of these two particles can be assumed collinear and perpendicular to the approach direction because the actual angular location of the point of closest approach ($180-\eta$) is dependent upon the eccentricity of the flyby orbit (Eq. [3.1]). With a minimum Leonid eccentricity of 79.405, the corresponding angle $180-\eta$ becomes 89.278 deg, implying that the point of closest approach is nearly perpendicular to the antiradiant direction. For slower-moving showers, the angle will decrease from the perpendicular, but even assuming the GEO case of 26.5 km/sec approach velocity, the angle 180–η is still more than 85 deg. As the point of closest approach moves away from the Earth, the eccentricity will increase and the point of closest approach will become even more perpendicular. With such a small deviation from 90 deg, there will be no loss of generality if the assumption is made in this part of the analysis that the points of closest approach for all meteor particles that pass by the Earth are collinear and perpendicular to the approach vector. In a larger context, this implies that the free-space flux $\Phi_{free\text{-}space}$ of the meteors can be assumed constant until the point of closest approach is reached.

The separation between the two orbits $\delta r$ will expand, owing to the influence of the Earth to a similar differential $\delta d$ at the crossover point in the antiradiant direction (Fig. 3.4). The differential crossover distance can therefore be found as a function of the differential closest approach distance. Assuming the differential distances are small, the closest approach point in Eq. (3.6) can be expressed for the two orbits as:

$$r_{p1} = r_1 \qquad \text{and} \qquad r_{p2} = r_2 = r_1 + \delta r \tag{3.7}$$

Equation (3.7) can be substituted into Eq. (3.6) to find the differential crossover distance (where higher-order terms have been neglected) as:

$$\delta d = \left(1 + \frac{V_\infty^2}{\mu} r_1\right)\delta r = e_1 \delta r \tag{3.8}$$

However, this takes into account only a single radial spike for the closest-approach distance between two particles. It was originally desired to find the total flux along the antiradiant direction because that expression will describe the maximum flux in the region behind the Earth to which satellites will be subject because of the focusing effect. Therefore, instead of analyzing two particles separated by $\delta r$, consider a ring of particles with two closest-approach distances separated by the differential $\delta r$ that contain a known free-space flux $\Phi_{free\text{-}space}$ (Fig. 3.5).

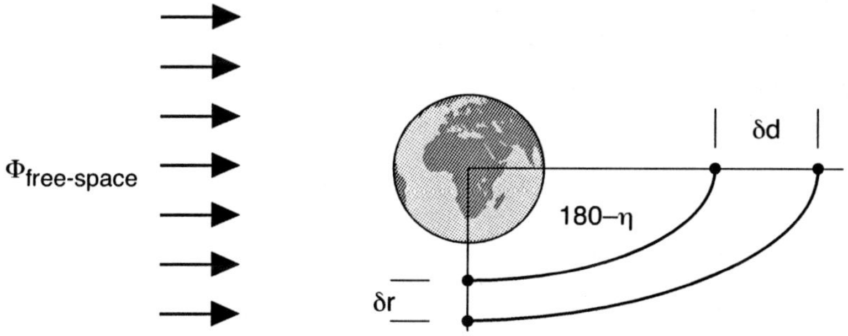


**Fig. 3.4. Close-approach geometry.**

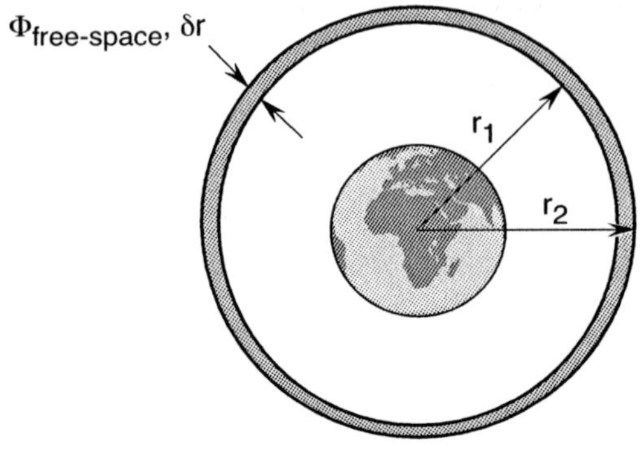


**Fig. 3.5. Cross-section of Earth flyby.**

To find the total number of particles per second, simply integrate over all the radial spikes:

$$N = 2\pi\Phi_{\text{free-space}} r_1 \delta r \tag{3.9}$$

Similarly, the number of particles per second occurring along the antiradiant vector will be a function of the flux along the antiradiant direction $\Phi_{\text{radiant}}$:

$$N = \Phi_{\text{radiant}} \delta d \tag{3.10}$$

It should be noted that the units of the antiradiant flux are not in particles/km$^2$/sec like the free-space flux. Because the antiradiant flux is determined along a one-dimensional line, the proper units are particles/km/sec.

Equating the two expressions for the number of particles and substituting Eq. (3.8), relating the differential crossover distance to the differential closest-approach distance gives the expression for the antiradiant flux as a function of the free-space flux of the meteors:

$$\frac{\Phi_{\text{radiant}}}{\Phi_{\text{free-space}}} = \frac{2\pi r_p}{1 + \frac{V_\infty^2}{\mu} r_p} = \frac{2\pi}{e_c} r_p \tag{3.11}$$

where $r_p$ is an arbitrary distance of closest approach (perigee of the flyby orbit) and where $e_c$ is the eccentricity of the flyby orbit corresponding to that particular distance of closest approach (Eq. [3.5]).

For the case of the Leonids, this gives rise to an antiradiant flux of about 505 times the free-space flux in the vicinity of the initial crossover point. From the minimum value at the initial crossover point, the antiradiant flux will increase until a limiting value of 511 is reached infinitely far from the Earth ($r_p$ goes to infinity in Eq. [3.11]):

$$\left(\frac{\Phi_{\text{radiant}}}{\Phi_{\text{free-space}}}\right)_{\text{lim}} = R_{\Phi-\text{lim}} = \frac{2\pi\mu}{V_\infty^2} \tag{3.12}$$

The increase in the flux ratio along the antiradiant therefore changes only rather slowly. Examining the limit then gives general insight into the problem. For the GEO case of a shower at 26.5 km/sec, the limit of the antiradiant flux is ~3500. It should be noted that the Draconids, a shower that in the past has produced extremely intense storms, approaches the Earth at ~23 km/sec. The limiting flux ratio along the antiradiant for the Draconids is ~4700 times the free-space value. Figure 3.6 shows the limiting value of the flux ratio along with the crossover distance for varying meteor speeds. For slower-moving meteors, the focusing can be extremely large.

### 3.2.2 Flux in the Near-Earth Environment

While this high particle density would seem to indicate that satellites would have a problem due to the focusing effect, this is not necessarily the case. This high flux value will occur only along the antiradiant vector since every differential ring (Fig. 3.5) will focus its entire particle content on a single point. Additionally, the Leonid crossover distance is so far from the vicinity of the Earth that virtually all satellites that are close by need not worry about the flux along the antiradiant direction. Of more crucial interest is the flux in the near-Earth environment.

Figure 3.7 shows the geometry for the region closer to the Earth than the crossover point. Note that a change in notation has occurred. The $x$ direction here denotes the perigee vector, which is again assumed collinear with all other flyby perigee vectors since the eccentricity is large for the flybys. The $y$ direction corresponds to the antiradiant vector considered earlier. Two orbits are examined that are in close proximity to each other throughout the flyby. This

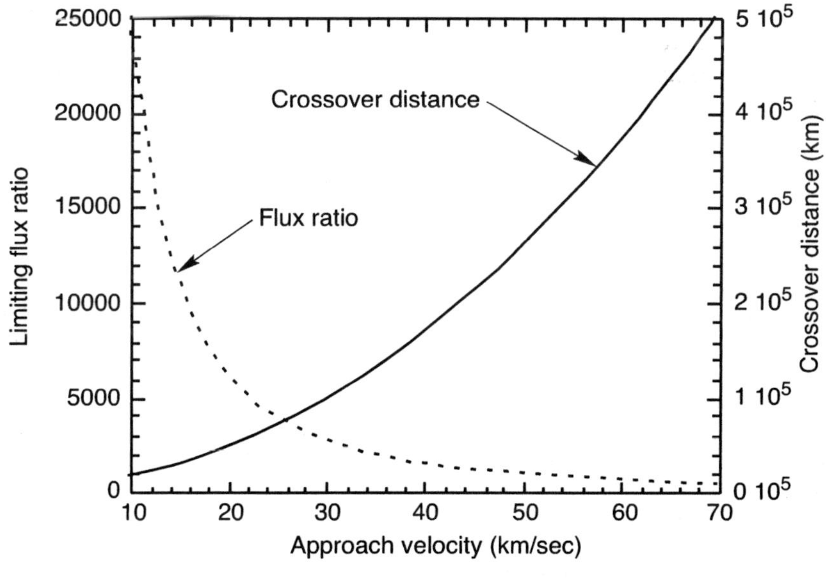


**Fig. 3.6. Limiting flux ratio and crossover distance for meteor flybys of the Earth.**

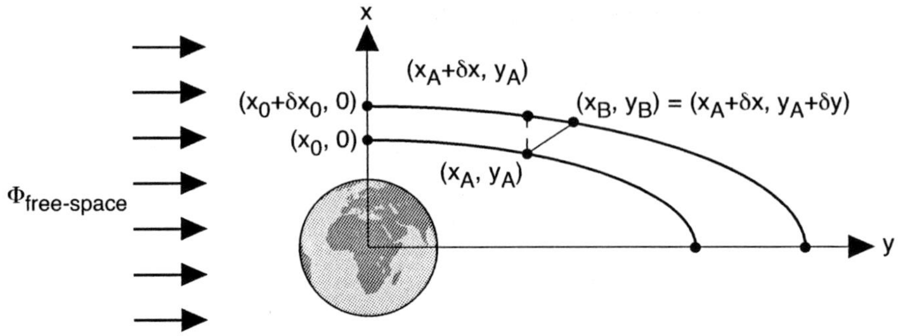


**Fig. 3.7. Flyby geometry at an intermediate midpoint.**

assumption is validated in Eq. (3.8), which shows that even if the particles are separated initially by 1 km in the $x$ direction, the particles will be crossing the antiradiant vector by the value of the flyby eccentricity. For a close Leonid flyby of the Earth, this means only 79 km of separation ($e = 79.405$). By comparing this with the minimum crossover distance itself of 256,000 km, it can be seen that the difference between the two orbits will always be small.

The number of particles per second will stay constant between the two orbits at equivalent times:

$$N = 2\pi\Phi_{\text{free-space}} x_0 \delta x_0 = 2\pi\Phi_A x_A \sqrt{\delta x^2 + \delta y^2} \tag{3.13}$$

The differentials $\delta x$ and $\delta y$ can be found from hyperbolic orbit geometry and the time of flight equation in the following manner. Because the orbit energy is the same between the two orbits, the semimajor axis of both flyby orbits is identical (mean motions are the same). Setting the time-of-flight expressions equal in Kepler's Equation for hyperbolic orbits yields:

$$n\Delta t = e_A \sinh F_A - F_A = e_B \sinh F_B - F_B \tag{3.14}$$

where $F$ is the hyperbolic anomaly at that point in the orbit.

Assuming the orbits are close together, orbit $B$ can be expanded about the values of orbit $A$:

$$e_B = e_A + \delta e \qquad F_B = F_A + \delta F \tag{3.15}$$

Substituting, an expression can be derived equating the differential change in the hyperbolic anomaly to the differential difference in the eccentricities of the two orbits:

$$\delta F = \frac{\sinh F_A}{1 - e_A \cosh F_A} \delta e \tag{3.16}$$

The orbit locations specific to each point along both orbits can be found by examining the hyperbolic geometry. The equations for the $x$ and $y$ coordinates are (Vallado, 1997):

$$x = a_\infty(\cosh F - e) \qquad y = -a_\infty \sqrt{e^2 - 1}\,\sinh F \tag{3.17}$$

The $\delta y$ between the two orbits can be found as:

$$\delta y = y_B - y_A = -a_\infty \sqrt{e_B^2 - 1}\sinh F_B + a_\infty \sqrt{e_A^2 - 1}\sinh F_A \tag{3.18}$$

By substituting the expansions for the hyperbolic anomaly and the eccentricity (Eq. [3.15]) along with Eq. (3.16) relating the hyperbolic anomaly and eccentricity differentials, $\delta y$ becomes:

$$\delta y = \frac{-a_\infty \sqrt{e_A^2 - 1}\sinh F_A}{1 - e_A \cosh F_A}(e_A - \cosh F_A)\frac{\delta e}{e_A^2 - 1} \tag{3.19}$$

The conic equation in the hyperbolic anomaly:

$$r = a(1 - e\cosh F) \tag{3.20}$$

can be used along with Eq. (3.17) to generate the final expression for $\delta y$:

$$\delta y = -\frac{x_A y_A}{r_A}\frac{\delta e}{e_A^2 - 1} \tag{3.21}$$

The differential $\delta x$ can be derived in a similar manner. The difference between the $x$ coordinates of the two orbits is:

$$\delta x = x_B - x_A = a(\cosh F_B - e_B) - a(\cosh F_A - e_A) \tag{3.22}$$

Expanding again in the hyperbolic anomaly and the eccentricity and then substituting for the conic equation and the $x$ and $y$ coordinates (Eqs. [3.15], [3.16], [3.17], and [3.20]), the differential becomes:

$$\delta x = \left[\frac{y_A{}^2}{r_A} + p_A\right]\frac{\delta e}{e_A^2 - 1} \tag{3.23}$$

Comparing Eqs. (3.21) and (3.23), where the $x$ and $y$ coordinates and the radius $r$ (Eq. [3.2]) are written in terms of the true anomaly, $f$:

$$x = r\cos f \qquad\qquad y = r\sin f \tag{3.24}$$

the ratio of the $\delta x$ and $\delta y$ can be computed as:

$$\frac{\delta y}{\delta x} = \frac{-\sin f \cos f}{2 + e\cos f - \cos^2 f} \tag{3.25}$$

Because the eccentricity will always be large, the ratio of the differential distances will thus be bound, regardless of the value of the true anomaly as approximately:

$$\frac{\delta y}{\delta x} \approx \pm\frac{1}{e} \tag{3.26}$$

Thus, since the eccentricity can never be smaller than 79.405 for the Leonids (and 12.237 for the GEO-shower case), the additional assumption that $\delta y$ can be ignored is allowable. Equation (3.13) relating the fluxes becomes:

$$\Phi_{\text{free-space}} x_0 \delta x_0 = \Phi_A x_A \delta x \tag{3.27}$$

By substituting Eq. (3.23) along with the eccentricity differential found by expanding Eq. (3.4):

$$\delta e = \frac{V_\infty^2}{\mu} \delta x_0 \tag{3.28}$$

into Eq. (3.27), the relation between the fluxes at any intermediate point and the free-space flux can be found:

$$\Phi_{\text{free-space}} x_0 = \Phi_A x_A \left[ \frac{y_A^2}{\sqrt{x_A^2 + y_A^2}} + p_A \right] \frac{V_\infty^2}{\mu(e_A^2 - 1)} \tag{3.29}$$

The general equation describing a hyperbola can be used to express Eq. (3.29) solely in terms of the $y$ coordinate (the distance from the center of the Earth along the antiradiant direction) (Roy, 1978):

$$\frac{(x + a_\infty e)^2}{a_\infty^2} - \frac{y^2}{a_\infty^2 (e^2 - 1)} = 1 \tag{3.30}$$

However, this expression assumes that the perigee is directly along the $x$ axis. In fact, the hyperbola will be offset by an angle equal to $90 - \eta$. Recall earlier that the assumption was made that $\eta$ was close enough to 90 deg to warrant taking the position that the perigee vectors were collinear and that the flux across two close orbits would be equivalent to the free-space flux. In using the equation for the hyperbola, this assumption cannot be made. Because of the shallowness of the orbit (high eccentricity), if the $x$ coordinate in Eq. (3.30) is taken as zero, the $y$ coordinate gives as a result the value of the semilatus rectum $p$. From Eq. (3.4), however, it is known that the proper $y$ coordinate in this case is the crossover point equal to one-half of the semilatus rectum. Close to the Earth the difference will be minuscule, but far away when the crossover point is approached, a factor of 2 cannot be ignored. As a consequence, the hyperbola of Eq. (3.30) must be rotated through the angle $90–\eta$ in order to have the correct geometric relation between $x_A$ and $y_A$. This relation is (Thomas, 1980):

$$(e^2 - 1)x^2 + 2exy + 2a_\infty e(e^2 - 1)x + 2a_\infty(e^2 - 1)y + a_\infty^2(e^2 - 1)^2 = 0 \tag{3.31}$$

Separating the variables yields:

$$y_A = -\frac{a_\infty^2 (e^2 - 1)^2 + (e^2 - 1)x_A^2 + 2a_\infty e(e^2 - 1)x_A}{2a_\infty(e^2 - 1) + 2ex_A} \tag{3.32}$$

The value $x_A$ can be replaced by the closest-approach distance $r_p$ used earlier, and $y_A$ is again equivalent to the antiradiant coordinate $d$. Recall also that the semimajor axis $a_\infty$ is negative for hyperbolic orbits and will be constant for all the flybys:

$$a_\infty = -\frac{\mu}{V_\infty^2} \tag{3.33}$$

Together with Eq. (3.5) rewritten in terms of the semimajor axis:

$$e = 1 - \frac{r_p}{a_\infty} \tag{3.34}$$

Equations (3.29) and (3.32) now give the flux ratio at any point after Earth encounter in terms of the free-stream conditions, the distance of closest approach for that particle's trajectory, and the antiradiant distance. The final expressions are:

$$\frac{\Phi_A}{\Phi_{\text{free-space}}} = \frac{r_p}{x_A}\left[\frac{y_A^2}{\sqrt{x_A^2 + y_A^2}}\frac{a_\infty}{r_p(2a_\infty - r_p)} + 1\right]^{-1}$$

and (3.35)

$$y_A = -\frac{r_p - 2a_\infty}{2}\left(\frac{r_p}{a_\infty}\right)\left[\frac{x_A{}^2 + 2(a_\infty - r_p)x_A + r_p(r_p - 2a_\infty)}{r_p(r_p - 2a_\infty) + (a_\infty - r_p)x_A}\right]$$

The closest-approach distance can never be less than the Earth's radius; $x_A$ will always be less than $r_p$. Similarly, the magnitude of the semimajor axis of the flyby orbits will also be much smaller than $r_p$ for all but the very slowest of meteor showers (Eq. [3.33]). Also, recall that the semimajor axis is negative for hyperbolic orbits. Making these assumptions simplifies Eq. (3.35) into:

$$\frac{\Phi_A}{\Phi_{\text{free-space}}} = \frac{3r_p}{2x_A} - \frac{1}{2} = \frac{3}{2}\left(\frac{r_p^2}{r_p^2 + 2a_\infty y_A}\right) - \frac{1}{2}$$

and (3.36)

$$y_A = -\frac{(r_p - x_A)}{2}\left(\frac{r_p}{a_\infty}\right)$$

Figure 3.8 shows the variation of the distance from the center of the Earth in the perpendicular direction versus the antiradiant distance for a particle that just misses the Earth's surface ($r_p$ = 6378 km) for the Leonids, the GEO shower, and the even slower Draconids (~23 km/sec). In essence, this establishes envelopes wherein Earth-orbiting satellites would experience some level of shadowing protection. The near-Earth environment is of interest in satellite operations, and for this region the shadow changes only slightly from a cylindrical shape for faster-moving showers. For the slower meteoroids, the shadow is more appropriately treated as conical.

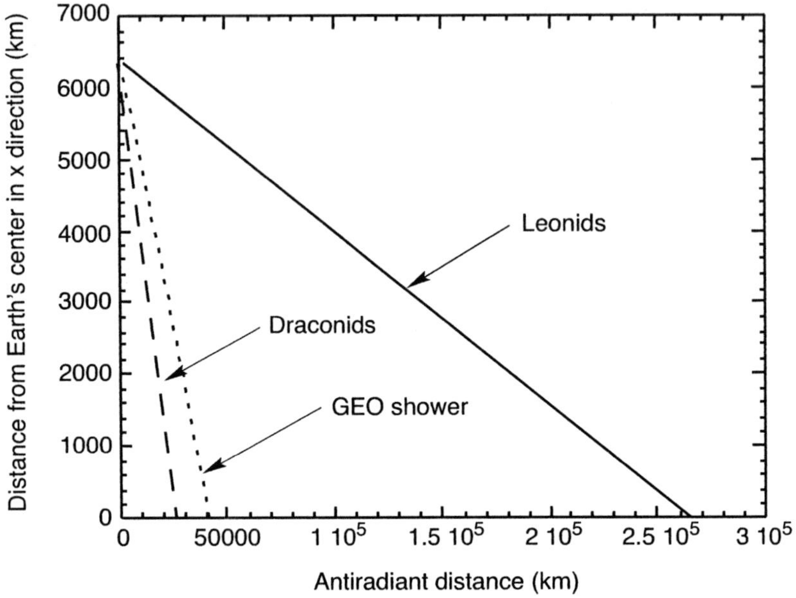


**Fig. 3.8. Envelope of shadow protection.**

Figure 3.9 shows the flux ratio for the same three examples. It can be seen that the flux ratio approaches infinity as the crossover point is reached. Again, in this particular situation, the linearization assumptions made in the analysis break down, and it is more appropriate to look at the flux along the antiradiant vector in the crossover region in order to gain insight into the problem. For near-Earth regions, the maximum orbital distance will never be greater than 50,000 km for a super-GEO type orbit. For this case, the Leonid flux increase because of the focusing of the Earth will be less than 10% for any closest-approach distance. This value is within the uncertainty of the actual fluxes for the Leonid showers; therefore, the focusing effect need not be considered in actual analysis of the satellite threat posed by these meteors. However, for the slower-moving showers, the focusing cannot be ignored because local values of the flux change rapidly with small changes in altitude.

### 3.2.3 Flux in the Crossover Region

While this describes the flux before crossover is reached, what happens afterward? In the * region of Fig. 3.3, the particles are sensing the effects of the Earth's mass but have not yet coalesced into a single point along the antiradiant vector. In the ** region of the same figure, some of the particles have crossed the antiradiant vector and are beginning to move away from each other. However, these particles will intersect with other meteors that have not yet crossed the antiradiant (point $A'$, Fig. 3.10).

Equation 3.35 describes the flux for the region before the antiradiant crossover occurs, but this does not completely describe the total flux after crossover. Two effects need to be considered: focusing the particle orbits that have not yet crossed

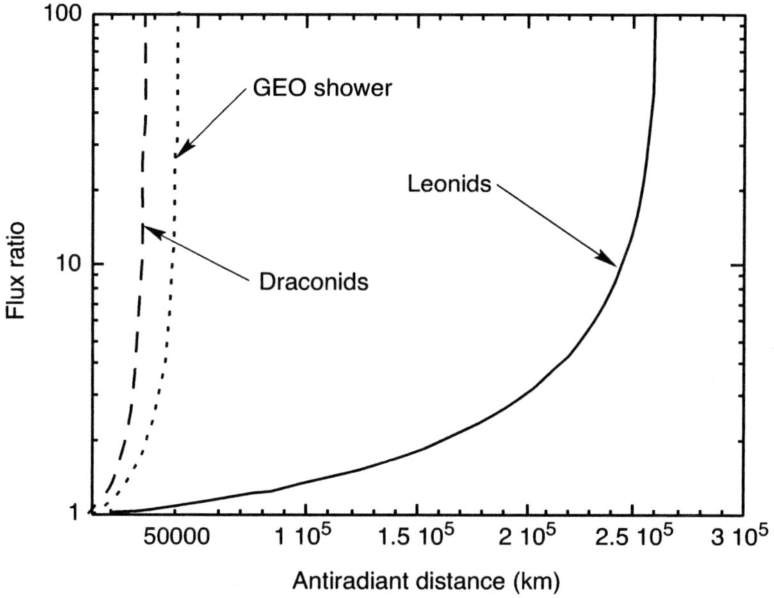


**Fig. 3.9. Flux ratio in the near-Earth environment.**

the antiradiant vector and those orbits that have already crossed it. The particle flux at point $A'$ after crossover will be the total of the two separate fluxes:

$$\Phi_{A'} = \Phi_A + \Phi_B \tag{3.37}$$

where $A'$ denotes the total particle flux at point $A'$, $A$ is the particle flux focused by those orbits that have not yet passed the antiradiant vector, and $B$ is the particle flux focused (and hence re-expanding) by those orbits that have crossed the antiradiant vector. The expressions for each of the fluxes are given by Eq. (3.36) written for those particular orbits. Summing the two components, the total flux ratio at $A'$ then becomes (Eq. [3.38]):

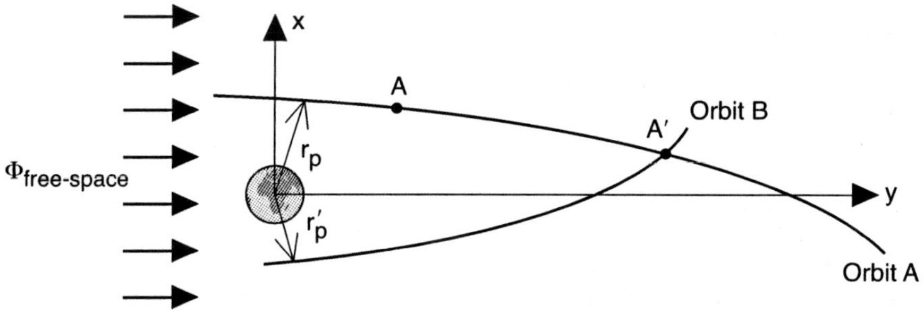


**Fig. 3.10. Intersection of particle fluxes in crossover region.**

$$\frac{\Phi_{A'}}{\Phi_{\text{free-space}}} = \frac{\Phi_A + \Phi_B}{\Phi_{\text{free-space}}} = \frac{r_{pA}}{|x_A|}\left[\frac{y_A^2}{\sqrt{x_A^2 + y_A^2}\, r_{pA}} \frac{a_\infty}{(2a_\infty - r_{pA})} + 1\right]^{-1} + \frac{r_{pB}}{|x_B|}\left[\frac{y_B^2}{\sqrt{x_B^2 + y_B^2}\, r_{pB}} \frac{a_\infty}{(2a_\infty - r_{pB})} + 1\right]^{-1} \quad (3.38)$$

where

$$y_A = -\frac{(r_{pA} - 2a_\infty)}{2}\left(\frac{r_{pA}}{a_\infty}\right)\left[\frac{x_A^2 - 2(a_\infty - r_{pA})x_A + r_{pA}(r_{pA} - 2a_\infty)}{r_{pA}(r_{pA} - 2a_\infty) - (a_\infty - r_{pA})x_A}\right] \quad (3.39)$$

and

$$y_B = -\frac{(r_{pB} - 2a_\infty)}{2}\left(\frac{r_{pB}}{a_\infty}\right)\left[\frac{x_B^2 - 2(a_\infty - r_{pB})x_B + r_{pB}(r_{pB} - 2a_\infty)}{r_{pB}(r_{pB} - 2a_\infty) - (a_\infty - r_{pB})x_B}\right] \quad (3.40)$$

The two orbits for which the fluxes are to be added will have different close-approach distances $r_{pA}$ and $r_{pB}$, opposite $x$ values ($x_A = -x_B$), but the same $y$ coordinate ($y_A = y_B$).

Figure 3.11 shows the large-scale view of the Leonid flux field for the Earth-Moon environment. The value plotted in this expression is the base10 log of the flux ratio plus unity and is simply an expression chosen for display purposes. In the figure, the area of complete shadow protection is zero (the black area), and the free-space flux is equal to unity. The crossover of the antiradiant vector begins at about 256,000 km from the Earth and is shown in the white area on the plot. Before the initial crossover point is reached, the particle flux changes very slowly with respect to the free-space flux, increasing to around three times the free-space value at a distance of 200,000 km (1.5 contour line).

Figure 3.12 shows a blowup of the crossover region. The contour value of 2.0 denotes a local flux equal to 10 times the free-space flux; the 3.0 contour is 100 times the free-space value. The single-order-of-magnitude-larger local flux is limited to a region of about 1000 km away from the antiradiant vector, while the 3.0 contour is about 95 km from the antiradiant. For a sample satellite at a circular orbit at this altitude (~300,000 km) traversing the maximum area, these distances correspond to a maximum amount of time spent in this regime of approximately 3 min for the region bounded by the 3.0 contour and 30 min for the 2.0 contour area. These times are well within the error on current knowledge of the time of the expected Leonid storm.

For a given point in the field ($x$,$y$), the closest-approach distance must be iteratively solved for. This type of numerical analysis using Eqs. (3.38) through (3.40) can be cumbersome if an initial approximate analysis is desired. In the environment close to the antiradiant vector under examination wherein the particle fluxes are approaching the focal point, several simplifying assumptions can be made.

The closest approach point (perigee of the flyby orbit $r_p$) will be large compared with both the $x$ axis (perpendicular distance from the antiradiant vector) and the semimajor axis of the flyby orbit (~ –81 km for the Leonids). In addition

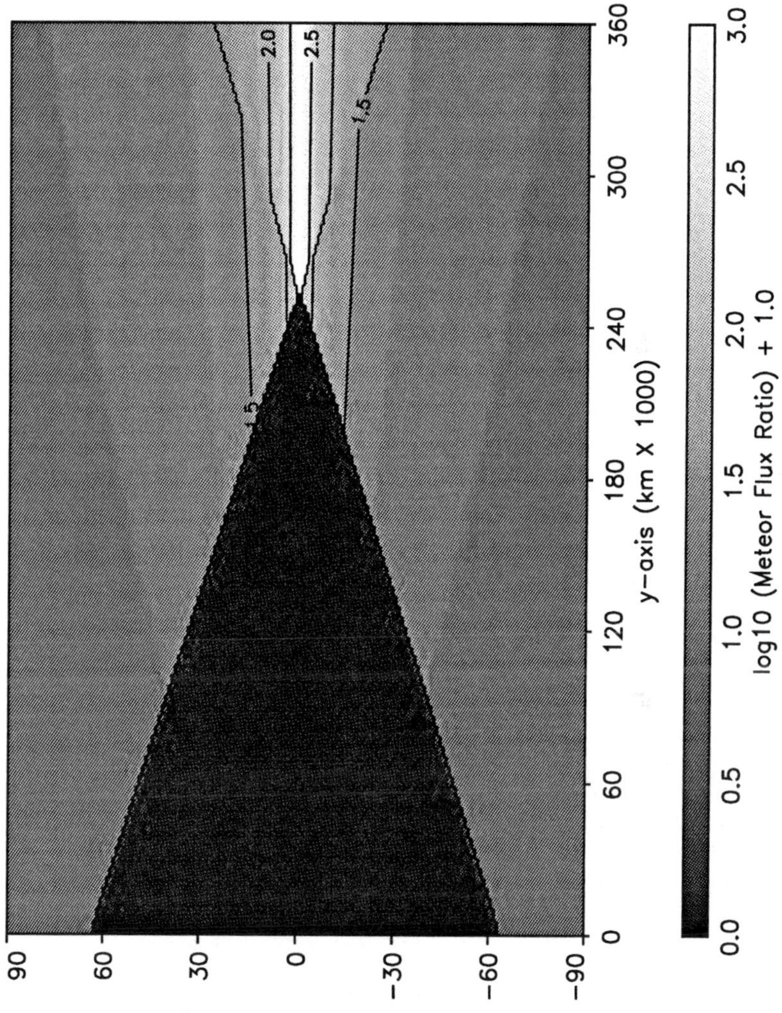


**Fig. 3.11. Large-scale view of flux field.**

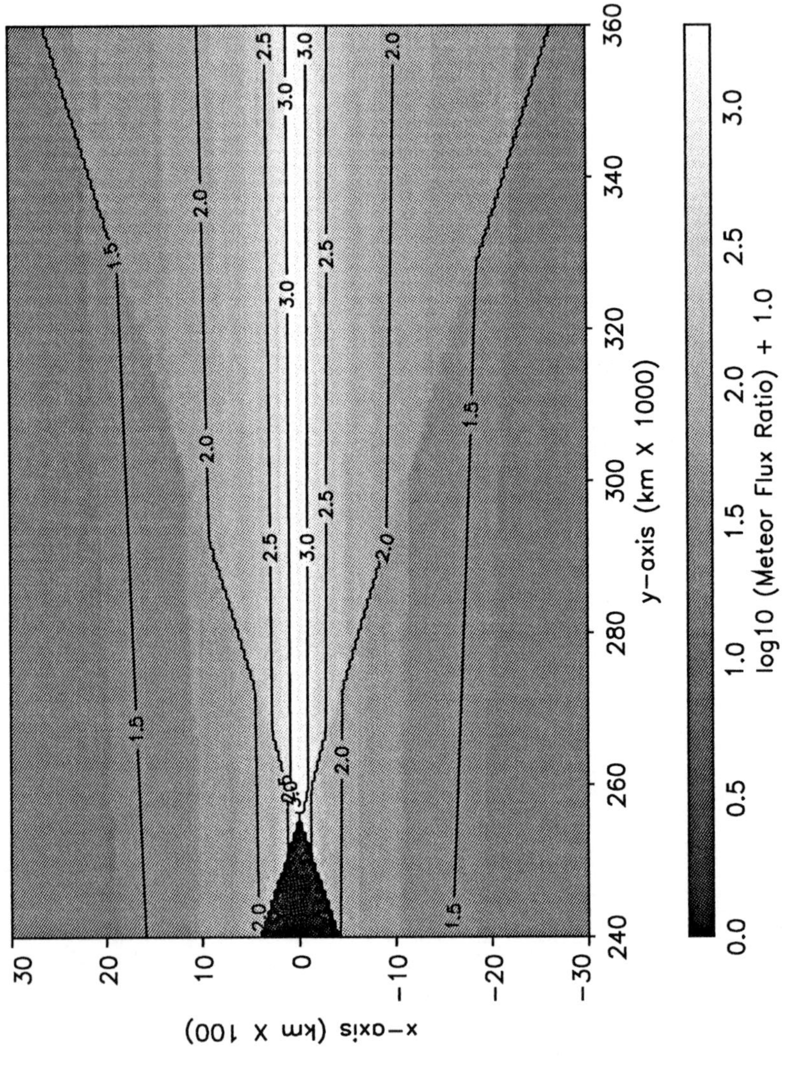


**Fig. 3.12. Small-scale view of flux field.**

because the initial closest-approach distance will be large compared with the $x$-axis distance, the higher-order terms of Eqs. (3.39) through (3.40) can be eliminated. The $y$-axis component then reduces to a simple expression in terms of the closest approach distance:

$$y_A = y_B = -\frac{r_p^2}{2a_\infty} \tag{3.41}$$

Substituting into Eq. (3.38) yields for the flux ratio close to the antiradiant vector:

$$\frac{\Phi_{A'}}{\Phi_{\text{free-space}}} \approx \frac{4}{3}\frac{\sqrt{-2a_\infty y_A}}{x_A} \tag{3.42}$$

Figure 3.13 shows the approximate equation results in the same units and scale as plotted in Fig. 3.12. It can therefore be concluded that in the near-radiant region, the approximate equation results give a good estimate of the particle flux caused by focusing out to approximately the 1000-km range for the Leonids.

Equation (3.42) can be generalized for any given meteor shower. Knowing the relation between the semimajor axis and the free-stream velocity of the meteors ($V_\infty$) passing by an arbitrary planet of gravitational constant $\mu$:

$$a_\infty = -\frac{\mu}{V_\infty^2} \tag{3.43}$$

Equation (3.42) can be reduced to:

$$\frac{\Phi_{A'}}{\Phi_{\text{free-space}}} \approx \frac{4}{3}\sqrt{2}\frac{\sqrt{\mu y}}{V_\infty x} \tag{3.44}$$

which is valid for any given meteor shower passing in the vicinity of any astronomical body. However, some caution is warranted; the approximate equation is good only when the distance $y$ is greater than the initial crossover point along the antiradiant vector and should not be used until that point has been reached.

In general, the time duration of a satellite in a near-circular orbit at an altitude where it will go through this dangerous regime can be calculated as well. The velocity of the spacecraft can be approximated by:

$$v_{s/c} = \sqrt{\frac{\mu}{y}} \tag{3.45}$$

and the distance traveled in an arbitrary region under examination (that passes directly through the antiradiant) is just twice the distance under which the flux ratio is chosen to be at a specifically desired level:

$$2x = \frac{8}{3}\sqrt{2}\frac{\sqrt{\mu y}}{V_\infty}\left(\frac{\Phi_{A'}}{\Phi_{\text{free-space}}}\right)^{-1} \tag{3.46}$$

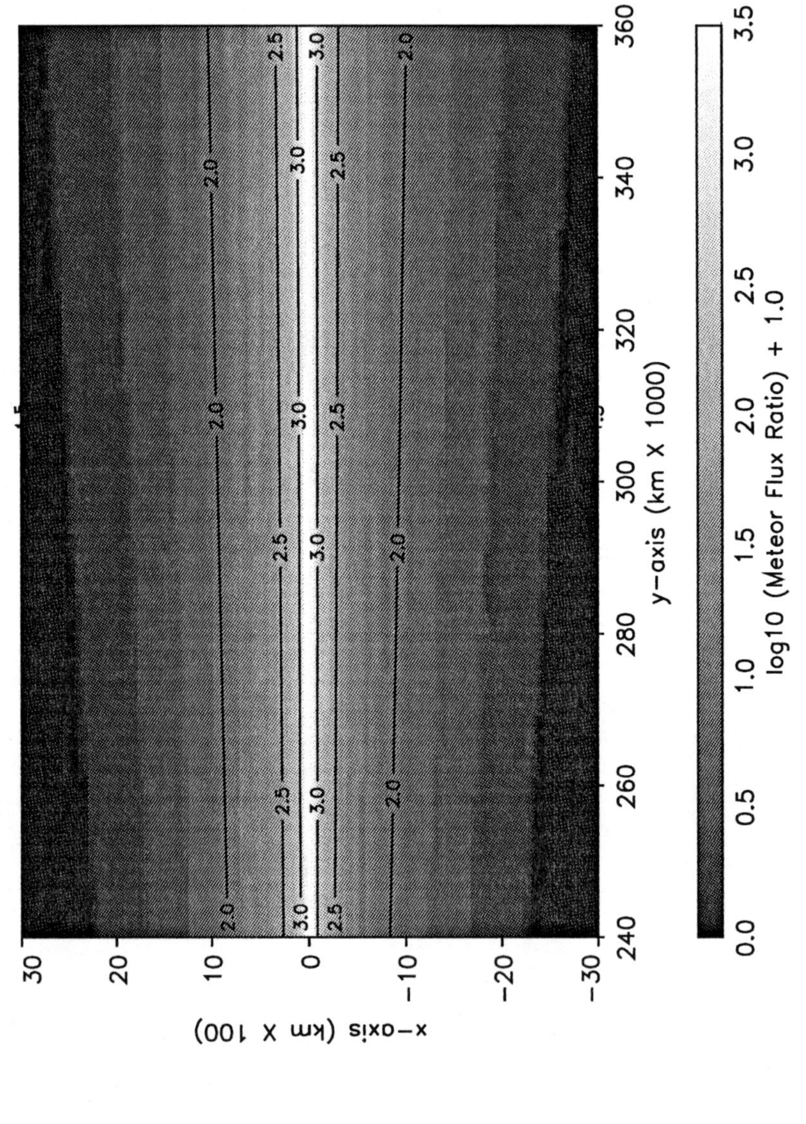


**Fig. 3.13. Approximate equation particle flux in the near-radiant environment.**

The time spent in this regime becomes:

$$t_t = \frac{2x}{v_{s/c}} = \frac{8}{3}\sqrt{2}\frac{y}{V_\infty}\left(\frac{\Phi_{A'}}{\Phi_{\text{free-space}}}\right)^{-1} \tag{3.47}$$

## 3.3 Integrated Flux: Actual Threat to the Spacecraft

Although Eq. (3.47) describes the total time spent in potential danger, it does not accurately depict the danger itself. It is more appropriate to look at the flux integrated over the crossover region and compare it with the free-space integrated flux that the spacecraft will sense as a result of the total event. Assuming symmetry about the antiradiant vector and where $t$ is one-half the total time of Eq. (3.47), the integrated flux (the fluence) is simply:

$$\Phi_I = 2\int_0^t \Phi_{A'}\delta t \tag{3.48}$$

Substituting Eq. (3.46) and utilizing the differential distance instead of performing the integration in the time domain gives:

$$\Phi_I = 2\int_0^x \frac{4}{3}\sqrt{2}\frac{\sqrt{\mu y}}{V_\infty v_{s/c}}\Phi_{\text{free-space}}\frac{\delta x}{x} \tag{3.49}$$

However, this expression contains a singularity along the antiradiant. To take this into account, it must be recalled that the particle flux along the antiradiant vector was approximately 500 times the free-space flux for the Leonids. Therefore the substitution of Eq. (3.44) into Eq. (3.48) should be conducted only until this flux ratio of 500 is reached. The $x$ coordinate corresponding to this level of flux ratio can be generalized by using the limiting value of the antiradiant flux ratio from Eq. (3.12), resulting in:

$$x_L = \frac{2V_\infty}{3\pi}\sqrt{\frac{2y}{\mu}} \tag{3.50}$$

For an antiradiant distance of 300,000 km, this gives an $x_L$ of about 17 km for the Leonids. Because the antiradiant flux changes only very slowly (Fig. 3.6), the flux ratio can be assumed at a constant level of the limiting flux ratio through the direct near-radiant environment from zero to $x_L$. The integrated flux then becomes:

$$\Phi_I = 2\int_{x_L}^x \frac{4}{3}\sqrt{2}\frac{\sqrt{\mu y}}{V_\infty v_{s/c}}\Phi_{\text{free-space}}\frac{\delta x}{x} + 2\int_0^{x_L} R_{\Phi-\text{lim}}\Phi_{\text{free-space}}\frac{\delta x}{v_{s/c}} \tag{3.51}$$

where $R_{\Phi-\text{lim}}$ is given by Eq. (3.12).

Evaluating and noting that if the distance $x$ is less than $x_L$, the first term must be taken directly as zero and the second integrated only to $x$; if the distance is greater than $x_L$, the second term should be integrated only to $x_L$:

$$\Phi_I = \frac{8}{3}\sqrt{2}\frac{y}{V_\infty}[ln x - ln x_L + 1]\Phi_{\text{free-space}} \qquad \text{if } x \geq x_L$$

$$\Phi_I = 2R_{\Phi-\lim}\sqrt{\frac{y}{\mu}x}\ \Phi_{\text{free-space}} \qquad \text{if } x < x_L \tag{3.52}$$

Equation (3.52) can be reformed in the time domain using Eq. (3.47) as:

$$\Phi_I = \frac{8}{3}\sqrt{2}\frac{y}{V_\infty}[ln(v_{s/c}t) - ln x_L + 1]\ \Phi_{\text{free-space}} \qquad \text{if } t \geq t_L$$

$$\Phi_I = 2R_{\Phi-\lim}\sqrt{\frac{y}{\mu}(v_{s/c}t)}\ \Phi_{\text{free-space}} \qquad \text{if } t < t_L \tag{3.53}$$

where $t_L$ is the time equivalent of $x_L$:

$$t_L = \frac{x_L}{v_{s/c}} \tag{3.54}$$

To gain greater insight into the relative danger that the flux concentration in the near-radiant environment poses to satellites, Eq. (3.53) can be compared to a satellite that spends an equivalent amount of time in the shower but away from the antiradiant. The integrated flux for that type of space vehicle is (note the time domain is chosen as one-half the total amount of time in the shower in order to be consistent with Eq. [3.53]):

$$\Phi_{I\text{-free-space}} = 2\int_0^t \Phi_{\text{free-space}}\ \delta t = 2t\Phi_{\text{free-space}} \tag{3.55}$$

The integrated flux ratio between the two spacecraft then becomes Eq. (3.53) divided by Eq. (3.55) and where Eq. (3.45) has been substituted:

$$\frac{\Phi_{I-\text{near-radiant}}}{\Phi_{I\text{-free-space}}} = \frac{8}{3}\sqrt{2}\frac{y}{V_\infty t_t}\left[ln\left(\frac{v_{s/c}t_t}{2}\right) - ln x_L + 1\right] \qquad \text{if } t_t \geq 2^* t_L$$

$$\frac{\Phi_{I-\text{near-radiant}}}{\Phi_{I\text{-free-space}}} = R_{\Phi-\lim} \qquad \text{if } t_t < 2^* t_L \tag{3.56}$$

where $t_t$ is the total amount of time of the interval. Recall that Eq. (3.55) was derived integrating over one-half of the time interval of interest.

Plotted in Fig. 3.14 for two antiradiant distances for the Leonids ($V_\infty = 70$ km/s), Eq. (3.56) (labeled "appr" in the figure) assumes that one of the spacecraft goes directly through the antiradiant vector and as such represents the worst case possible, while the other is completely separate from the focusing effect and therefore shows the best possible free-space case. In essence, the second vehicle is on the side of the Earth toward the direction from which the meteors are coming and therefore sees no focusing effect at all. Note that the simplifying assumptions used in deriving Eq. (3.56) are good only in the vicinity of the antiradiant.

As the time interval grows, the approximate expressions will become less and less valid. However, the qualitative behavior of the plot remains true. As the time interval under consideration grows to infinity, the integrated flux ratio will approach, but never quite reach, a limiting value close to unity.

The curves labeled "num" in Fig. 3.14 correspond to the full numerical solution. Good agreement is seen early on with the approximate equations. Because the satellite is moving more slowly at higher distances, the approximations made earlier become valid for longer and longer times. The abrupt deviation in the numerical $y = 300{,}000$ km curve from the corresponding approximate curve is the result of the satellite leaving the crossover region (** from Fig. 3.3) and entering the * region. A similar kink occurs for other cases examined ($y =$ 400,000 and 500,000 km), although not seen here for purposes of clarity and scale. It is at this point when the satellite leaves the crossover region that the approximate equations truly begin to break down.

It is anticipated that the most intense period of peak activity for the Leonids will last approximately 2 hours. Even though the focusing will exist whether or not a storm occurs, the focusing during an anticipated storm will generate the

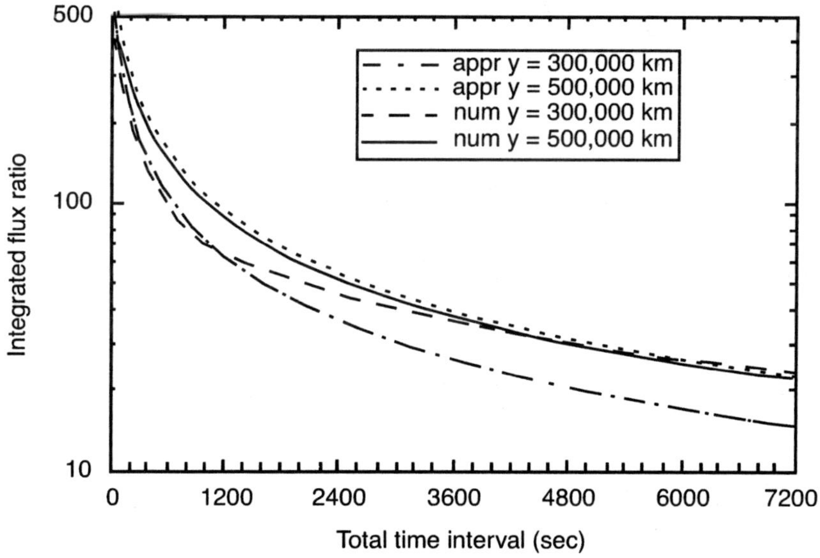


**Fig. 3.14. Integrated flux ratio comparing antiradiant crossing spacecraft with spacecraft far away from focusing effect.**

absolute worst conditions possible for orbiting spacecraft. If Eq. (3.56) is examined for a time span of this duration, Fig. 3.15 is generated from both the approximate equations and full numerical analysis. The deviation at lower altitudes is again the result of the spacecraft leaving the area of intersection and entering the region where the approximate equations are no longer valid.

For a 2-hour storm, the increase in the threat for a satellite traveling directly through the region of maximum intensity along the antiradiant as compared with a satellite experiencing no focusing effect of the meteors is over an order of magnitude and does not change greatly for large y distances. Note that this is for a 2-hour Leonid storm; if storms of other durations are expected, a separate figure will need to be generated. However, it is safe to conclude from Fig. 3.15 that regardless of the storm duration, an increase in the threat posed to a spacecraft of over an order of magnitude can be readily anticipated when traveling through the near-radiant environment.

This result can be expanded to include the full Leonid shower duration and not just the short period of intense activity coming from a 2-hour storm, ultimately yielding the limiting value of the danger. The flux integrated over a full orbital period can be computed for both a satellite that travels through the antiradiant and a spacecraft that spends the total period in the free-space environment. An appropriate time can be chosen that will dictate whether the spacecraft is in the near-radiant environment as opposed to the free-space conditions. For purposes of convenience (see Figs. 3.14 and 3.15), 7200 sec is a reasonable choice that will include the most intense region around the antiradiant vector for a wide range of $y$ values. The approximate limiting expression then becomes:

$$\left(\frac{\Phi_{I-\text{near}-\text{radiant}}}{\Phi_{I-\text{free}-\text{space}}}\right)_{\text{lim}} = 1 - \frac{7200}{TP}\left[1 - \frac{\Phi_{I-\text{near}-\text{radiant}}}{\Phi_{I-\text{free}-\text{space}}}(\text{eval @ 7200 s})\right] \quad (3.57)$$

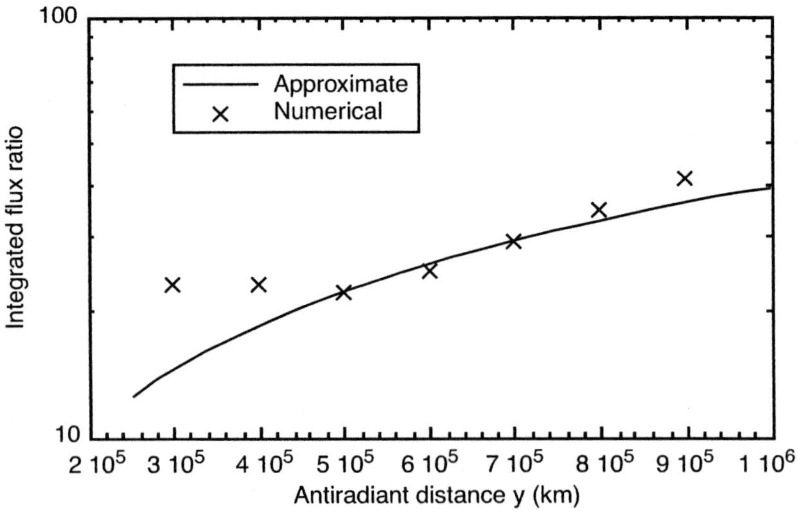


**Fig. 3.15. Integrated flux ratio for a 2-hour Leonid storm.**

where *TP* is the orbital period. Figure 3.16 then depicts the limiting values of the integrated flux ratio. It must be noted that this is only an approximate limit based upon the 7200 sec as being an interval that will contain the significant antiradiant flux focusing. Desired specific cases should be analyzed numerically for the full orbital period. It can be concluded, however, that for a Leonid shower (not storm) of several days' duration, the focusing will not have a significant impact on the overall threat to a space vehicle (<10%, as seen in Fig. 3.16). The small amount of time spent in the near-radiant environment will mostly average out during the entire orbital period. For the much shorter duration expected for the storm, satellites that are unfortunate enough to be in the antiradiant region when those particles pass by could experience over an order of magnitude greater threat than otherwise would be the case (Fig. 3.15). It should also be noted that since the near-radiant environment begins at 256,000 km and the particles are traveling at ~70 km/sec, just when the storm is about over for the Earth, it is just starting for the near-radiant satellites.

It should be emphasized that while the equations are valid for any meteor event that is encountered by the Earth, the specific numerical analysis presented applies only to the Leonids. Each desired meteor event should be examined separately. However, for a geosynchronous satellite that goes through a 2-hour Draconid storm ($V_\infty$ ~23 km/sec), a numerical analysis shows that the integrated flux ratio is ~10.1. Therefore a decent rule of thumb to consider is that while going through a focused region does not significantly increase the overall threat during a lengthy shower, a rough order of magnitude increase in the danger would be experienced by a satellite traversing the same region during a short-duration storm. The shorter the event, the greater the impact from the focusing. So not only does the storm increase the free-space flux and hence the threat to the spacecraft, but a focused storm will increase the danger by at least 10 times

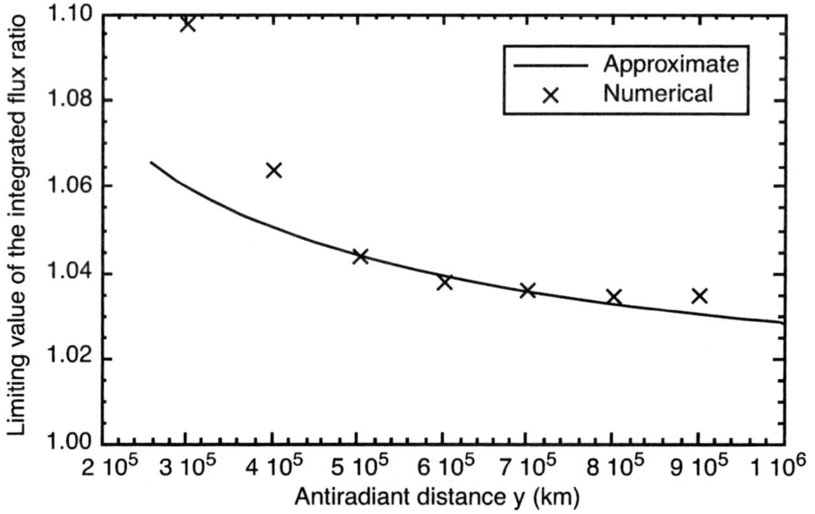


**Fig. 3.16. Integrated flux ratio for week-long Leonid shower.**

over that. Therefore the focusing effect should be kept in mind for any future meteor outbursts.

### 3.4 Lunar Focusing

The Earth is not the only body that can cause a focusing that would affect Earth-orbiting satellites. The Moon can as well. At first impression, the Moon would seem an unlikely candidate with its lesser mass and greater distance from the satellite. In addition, the unlikely alignment of the Earth-centered lunar location with the meteor radiant would be required. Figure 3.17 shows the antiradiant flux ratio and crossover distance for meteors flying past the Moon. With an Earth-Moon (and hence approximate satellite-to-Moon) distance of 384,400 km, only meteors approaching with a velocity of ~35 km/sec or less will affect Earth-orbiting vehicles (i.e., the Leonids at ~70 km/sec are not an issue for lunar focusing). Similarly, with a lunar-to-Earth mass ratio of 0.012, the antiradiant flux from lunar focusing will be only about 1% as intense as the focusing because of an Earth flyby (compare Fig. 3.17 with Fig. 3.6).

However, it is not the instantaneous flux along the antiradiant that poses the threat, but rather the integrated flux that the spacecraft flies through during the entire event (Eq. [3.56]). Figure 3.17 also shows the integrated flux ratios for an Earth-centered GEO satellite experiencing focusing from both the Earth and the Moon for a 2-hour outburst. The lunar values are larger than the Earth values by as much as an order of magnitude. Because the Earth-orbiting satellite is much further from the Moon than from the Earth, the crossover region caused by the

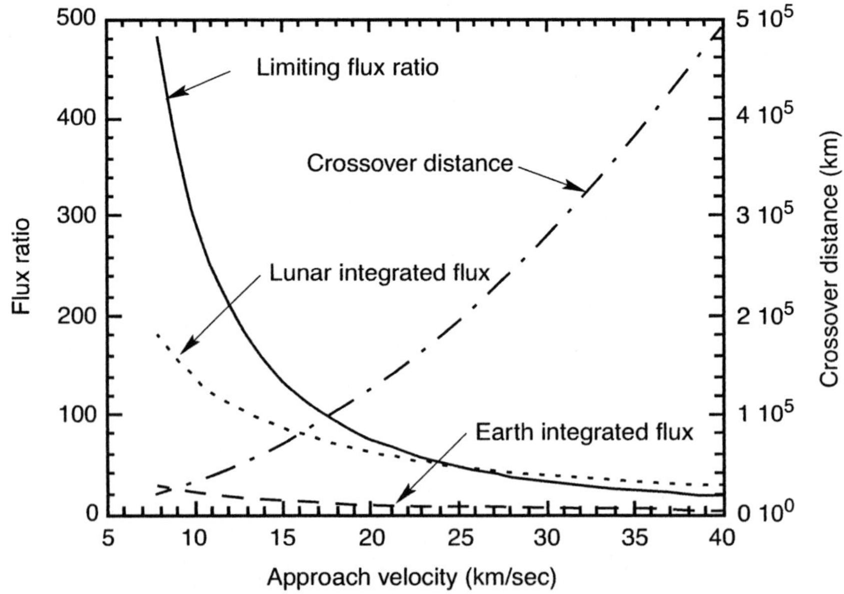


**Fig. 3.17. Limiting flux ratio and crossover distance for meteor flybys of the Moon.**

Moon that is experienced by the spacecraft is much larger than the crossover region created by the Earth. Thus, the greater integrated flux is created by the Moon because the satellite has to traverse a greater distance in the crossover region for the lunar focusing than for the Earth (** region in Fig. 3.3). This would seem to indicate that lunar focusing is a worse situation than Earth focusing, but one issue remains: alignment.

Consider Fig. 3.18. The amount of deviation from perfect alignment (i.e., Earth-Moon-meteor radiant are collinear) is going to be a function of the approach velocity. The angular distance Δ can be in either the right ascension, declination, or both. Its magnitude is driven by the perpendicular distance from the Moon-centered radiant to the furthest edge of the Earth-centered orbit. The angle Δ becomes:

$$\Delta = \frac{x + a}{R_{em}} \tag{3.58}$$

where $R_{em}$ is the Earth-Moon distance, and $a$ is the semimajor axis of the orbit. The distance $x$ is the perpendicular extent of the crossover region and is found from Eq. (3.46). For a lunar flyby, this gives:

$$\Delta = \left(\frac{8186}{V_\infty} + a\right)\frac{1}{R_{em}} \tag{3.59}$$

The first term in Eq. (3.59) will always be only a fraction of the second because $V_\infty$ will never be smaller than ~8 km/sec (Appendix B). The angular deviation from perfect collinear status can be approximated by the ratio of the semimajor axis to the Earth-Moon distance. For a GEO orbit, where the largest value can occur, the angular deviation allowable is thus ~6.3 deg. This implies that if the Moon's location is within about 6 to 7 deg (to allow for error) of the meteor radiant for a slow (<~35 km/sec) meteor event, operators should investigate the potential for lunar focusing.

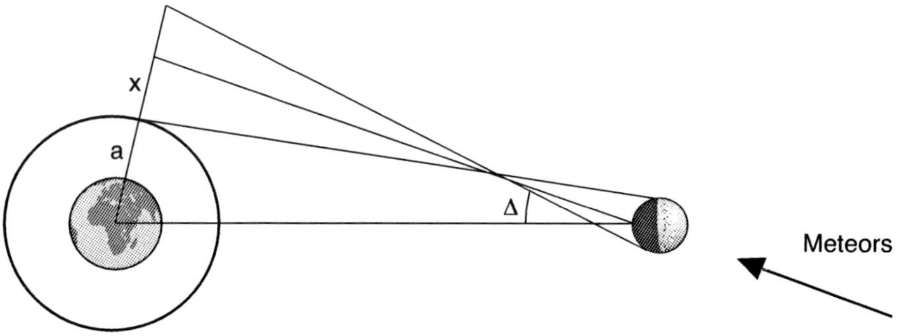


**Fig. 3.18. Lunar focusing geometry.**

While lunar focusing will therefore occur only rarely, when it does it could create a rather interesting display for observers on the Earth. Since the Earth, or at least a portion of it, is in the focused region, the meteors will appear to be coming from all different directions. In essence, the streaks will appear to weave and crisscross with abandon.

### 3.5 Summary of Focusing Effects

In general, the focusing caused by the gravitational field of the Earth is not a significant issue for long-duration meteor showers, but it can be for short-duration storms. The initial point where particles from opposite sides of the Earth will cross over is:

$$d = r_p + \frac{1}{2}\frac{V_\infty^2}{\mu} r_p{}^2 \tag{3.60}$$

and the most intense region of focused activity is small in size:

$$2x = \frac{8}{3}\sqrt{2}\frac{\sqrt{\mu y}}{V_\infty}\left(\frac{\Phi_{A'}}{\Phi_{\text{free-space}}}\right)^{-1} \tag{3.61}$$

where $x$ is the distance perpendicular from the antiradiant vector, $y$ is the distance along the antiradiant ($y$ must be greater than or equal to $d$ from Eq. [3.60]), and the flux ratio is chosen to be at whatever level is ascertained as being dangerous. For the Leonids with a local flux of 10 times the free-space value, $2x$ is less than 1700 km.

The focused flux along the antiradiant can have rather large values (~500 for the Leonids, ~4700 for the Draconids):

$$\left(\frac{\Phi_{\text{radiant}}}{\Phi_{\text{free-space}}}\right)_{\text{lim}} = \frac{2\pi\mu}{V_\infty^2} \tag{3.62}$$

The focused flux before crossover approximates as:

$$\frac{\Phi_A}{\Phi_{\text{free-space}}} = \frac{3}{2}\frac{r_p}{x_A} - \frac{1}{2}$$

and (3.63)

$$y_A = \frac{(r_p - x_A)}{2}\left(\frac{V_\infty^2 r_p}{\mu}\right)$$

Yet the actual threat to the spacecraft is not the instantaneous flux, but rather the integrated flux experienced over the course of the meteor event. Assuming the satellite goes directly through the antiradiant vector and comparing with an equivalent amount of time spent far away from the focused region, the integrated flux ratio is:

$$\frac{\Phi_I}{\Phi_{I-\text{free}-\text{space}}} = \frac{8}{3}\sqrt{2}\frac{y}{V_\infty t_t}\left[ln\left(\frac{v_{s/c}t_t}{2}\right) - lnx_L + 1\right] \qquad \text{if } t_t \geq \frac{2x_L}{v_{s/c}}$$

$$\frac{\Phi_I}{\Phi_{I-\text{free}-\text{space}}} = \left(\frac{\Phi_{\text{radiant}}}{\Phi_{\text{free}-\text{space}}}\right)_{\text{lim}} \qquad \text{if } t_t < \frac{2x_L}{v_{s/c}} \tag{3.64}$$

where $t_t$ is the total amount of time of the meteor event (either the shower or the storm) and $x_L$ is the value of the $x$ coordinate where the local instantaneous flux in the vicinity of the antiradiant vector reaches that value along the antiradiant:

$$x_L = \frac{2V_\infty}{3\pi}\sqrt{\frac{2y}{\mu}} \tag{3.65}$$

For the 2-hour Leonid storm, the values of the integrated flux can reach approximately 20 times the value of the free-space integrated flux; for the Draconids, about 10 times the free-space value. For either shower (as opposed to a storm), the integrated flux ratio approaches unity because the time spent in the antiradiant region is so much shorter than the total duration of the event.

In general, the shorter the duration of the meteor event and the slower the approach velocity of the particles, the more significant the effect of the gravitational focusing becomes. Short-duration storms pose a greater risk not only because of the storm but also because of the focusing; the focusing could also be significant for short-duration showers. However, meteoroids have to be travelling at less than ~27 km/sec to influence satellites at GEO altitudes or less. Longer-duration meteor events, while experiencing an identical focusing, will have the focusing average out over the longer duration, so this effect can be ignored for meteor showers several days in length.

This is also true for lunar focusing events. The Moon's smaller mass causes a shallower focusing, but the greater distance between Earth-orbiting satellites and the gravitating body results in a more intense focused integrated flux. However, the Moon must align to within 6 to 7 deg of the radiant point in order for lunar focusing to affect Earth-orbiting satellites.

# 4 Mitigating the Threat

Several strategies can be employed to dynamically mitigate the danger posed by any particular meteor event. The most obvious way of completely avoiding the Leonids or any other meteor threat is simply to not launch and thereby dodge the issue entirely. Of course, if the meteor events are annual or frequent, this tactic is not very effective. If, however, the event is singular, such as a meteor storm caused by the reappearance of the parent comet, then avoiding the threat by delaying the launch is indeed a viable option, provided more important mission constraints are not violated. But this is not a practical alternative for satellites already on orbit. Because past meteor events have been known to damage spacecraft to the extent that mission operations were significantly degraded or even completely lost, solutions to the on-orbit problem are most crucial.

One mitigation option is to alter the attitude of the spacecraft in order to minimize the cross-sectional area that is presented to the meteors, thereby reducing the probability of an impact. A related option is to orient the spacecraft so that the flux that penetrates a vehicle's shielding is minimized. This is potentially distinct from area minimization. Other ways of reducing the threat consist of in-orbit maneuvering to reduce the relative velocity between the spacecraft and the meteoroids, orbital maneuvering to place as much distance as possible between the spacecraft and the center of the meteor stream, and out-of-plane maneuvering to place the node in a position that takes advantage of the Earth's shadowing. Attitude control, in-plane maneuvers, and node control will be examined in detail in this chapter. Operational considerations are unique to each type of spacecraft and mission; practicalities may impact the applicability of the examined strategies. Therefore the studies will be presented in a generalized format.

### 4.1 Safing the Satellite

For on-orbit situations, the easiest and safest means of reducing the probability of damage is simply to safe the satellite to avoid unnecessary damage from a physically catastrophic strike from a large meteor or electromagnetic pulses (EMP) and plasma discharges arcing across the spacecraft resulting from a small meteor impact. Safing the vehicle also allows mission operators to assess potential damage before deciding on a recovery procedure. This removes the possibility of worsening the situation through an incorrect response to a meteor strike.

An example of the latter phenomenon occurred with the Olympus I telecommunications satellite, lost during the Perseid meteor shower in 1993. The vehicle's loss was not the result of a catastrophic type of impact; rather, it was largely the result of automatic procedures that went into place when the satellite was struck. Loss of pointing occurred with impact, and the vehicle subsequently went into its preplanned Sun reacquisition mode. By the time the vehicle's orientation was successfully reestablished, so much fuel had been used that the mission was effectively ended. This was the first known incident that involved loss of attitude control. Operators can learn from this unfortunate experience: safing the satellite during the expected time of the storm can reduce the likelihood of unnecessary operations that could negatively impact mission lifetime (Caswell, 1998).

## 4.2 Attitude Control for Damage Minimization

After safing the satellite, attitude control is the easiest and operationally safest method of reducing the potential of a meteor impact. One fundamental concept of attitude control is to minimize the cross-sectional area of the spacecraft in order to reduce the probability of impact. With a smaller area presented to the oncoming stream, the chances of a damaging strike are reduced. Attitude control for area minimization can be characterized in terms of three spacecraft-critical parameters: the vehicle (body of the spacecraft), the solar panels, and the sensors.

The other fundamental concept of attitude control is to not minimize the cross-sectional area, but rather to maximize the angle of the surface with respect to the spacecraft body surfaces so that the risk of particle penetration and spall production goes down. This is not applicable to those portions of a spacecraft such as a solar panel or a deployed antenna exposed directly to the meteoroid environment. But it does apply to the body of the spacecraft itself, which is likely to have shielding of some sort that protects the internal workings of the vehicle. The damage in the latter case does not come from the number of particles that hit the vehicle body, but rather from the number of particles that penetrate the shielding to the extent that spall or an electrostatic discharge (ESD) is produced. The objective of this strategy is to minimize the damage, not necessarily the number of strikes.

This section examines orienting the vehicle body away from the minimal cross-sectional area thereby potentially lessening the level of damage sustained by the body. It is shown that under a wide range of circumstances, including the Leonid meteor event, that this orientation is significantly better in terms of reducing the damage than the minimal cross-sectional area attitude.

### 4.2.1 Spall and ESD Damage

Although many types of shielding exist, the two most common are a single wall and a Whipple double wall. Another is the multishock shield, which reduces the weight requirements and produces less spall during impact. Other types of shielding also exist, and the analysis that follows will be general in nature to account for this variation. The ballistic limiting equations describing what it takes to penetrate a spacecraft's shielding and produce spall for the three cases are (Christiansen, 1993; Anderson and Cooke, 1998)

$$\text{single:} \qquad d_c = \left[\frac{t_w}{9.43}H^{1/4}\left(\frac{\rho_w}{\rho_p}\right)^{1/2}\left(\frac{V_n}{c}\right)^{-2/3}\right]^{18/19}$$

$$\text{double:} \qquad d_c = \left(\frac{t_w}{0.1289}\right)^{2/3}\rho_p^{-1/3}\rho_b^{-1/9}V_n^{-2/3}S^{1/3}\left(\frac{\sigma}{70}\right)^{1/3} \qquad (4.1)$$

$$\text{multishock:} \qquad d_c = 0.358t_w^{1/3}\left(\frac{\rho_w}{\rho_p}\right)^{1/3}V_n^{-1/3}S^{2/3}\left(\frac{\sigma}{40}\right)^{1/6}$$

where $t_w$ is the thickness of the wall (cm), $\rho_w$ is the density of the wall (g/cm$^3$), $\rho_b$ is the density of the buffer (g/cm$^3$), $H$ is the Brinnell Hardness, $c$ is the speed of sound (km/sec), $S$ is the bumper standoff distance (cm), $\sigma$ is the yield stress (ksi), $\rho_p$ is the particle density (g/cm$^3$), $d_c$ is the critical-particle diameter (cm), and $V_n$ is the component of the particle approach velocity that is normal to the wall (km/sec). Rewriting Eq. (4.1) in general terms gives

$$d_c = \left[\frac{G_S}{\rho_p^{1/2}(V_n)^{\gamma_S}}\right]^{18/19} \quad d_c = \left[\frac{G_D}{\rho_p^{1/2}(V_n)^{\gamma_D}}\right]^{2/3} \quad d_c = \left[\frac{G_{MS}}{\rho_p(V_n)^{\gamma_{MS}}}\right]^{1/3} \tag{4.2}$$

where $G_s$, $G_D$, and $G_{MS}$ represent combined parameters that are solely a function of the physical properties of the wall. The normal velocity is simply the relative velocity between the meteoroids and the spacecraft multiplied by the cosine of the angle between the meteor radiant and the vector normal to the wall:

$$V_n = v_{\text{rel}} \cos\vartheta \tag{4.3}$$

The single-wall parameter $\gamma_s$ varies between models and can have values ranging from 0.472 to 0.875, with a typical value of approximately two-thirds. Similarly, $\gamma_D$ and $\gamma_{MS}$ vary but are usually a value equal to or slightly less than unity. Therefore, whether considering a variety of shield types, the size of a meteoroid that will penetrate a given thickness can be approximated as

$$d_c \approx \frac{G_m}{(v_{\text{rel}} \cos\vartheta)^{\gamma_p}} \tag{4.4}$$

where $G_m$ is a parameter combining either $G_s$, $G_D$, or $G_{MS,}$ along with the particle's density and where the velocity index $\gamma_p$ is positive but less than unity.

It should be noted that the ballistic limiting equations [Eq. (4.1)] are empirically determined from impact velocities of less than 10 to 12 km/sec. As a consequence, whether the ballistic equations are applicable to meteoroids of Leonid velocity (which are at the extreme maximum of meteor velocities at ~71 km/sec) is unknown. For this reason, the parameter $\gamma_p$, while less than unity for the given ballistic equations, will be carried along in the analysis to emphasize that $\gamma_p$ is subject to change with better knowledge of the physics of hypervelocity impact and will vary for different shields. Additionally, the normal velocity dependence of Eq. (4.4) may not be valid for very shallow impact angles on the order of a few degrees (Ben-Schmuel, 1999). However, for the moderately acute angles that will be derived here, the relation will hold.

The benefit in carrying $\gamma_p$ along as an independent parameter becomes even more apparent when considering phenomena other than just shield penetration. An ESD can be produced by smaller particles that do not penetrate the shielding but that cause damage to a vehicle's electronics without generating mechanical damage. The formula relating the charge generated from a particle impacting a shield has the form (Campbell and Medina, 1998; Ratcliffe *et al.*, 1995)

$$Q = 8.54 \times 10^{-6} m_p^{\varepsilon} V_n^{\gamma_E - 1} \tag{4.5}$$

The impulsive plasma current associated with such a charge can also damage electrical systems. It is a simple matter to divide the charge by the characteristic time, $L/V_n$ (where $L$ is the plasma path length) to find the current. This effectively increases the incident velocity exponent by 1 (McDonnell *et al.*, 1997). The plasma current is then

$$I = G_E m_p^{\varepsilon} V_n^{\gamma_E} \tag{4.6}$$

where $G_E$ is a constant, $\gamma_E$ is an exponent ranging from 4 to 6 in different models, $\varepsilon$ is close to unity (Ratcliffe *et al.*, 1995; Campbell and Medina, 1998) and where the mass of the spherical particle is found as

$$m_p = \frac{4}{3}\pi\left(\frac{d_c}{2}\right)^3 \rho_p \tag{4.7}$$

Assuming that a certain level of electrical activity will damage the spacecraft electronics through an ESD, the minimum particle diameter can be solved for (where $G_I$ is yet another constant involving $G_E$, the limiting value of the damaging plasma current, and the physical properties of the meteoroid particles) as:

$$d_c = \frac{G_I}{(v_{\mathrm{rel}}\cos\vartheta)^{\gamma_E/3}} = \frac{G_I}{(v_{\mathrm{rel}}\cos\vartheta)^{\gamma_p}} \tag{4.8}$$

Hence, Eq. (4.4) can be used regardless of whether penetration of the shielding or damaging plasma production is being considered, as long as the difference in the parameter $\gamma_p$ is noted (i.e., $\gamma_p$ ~2/3 for physical penetration for single or Whipple shields, $\gamma_p$ ~ 1/3 for multishock shields, $\gamma_p$ ranges from ~4/3 to ~2 for plasma effects). By increasing the minimum particle diameter, $d_c$, which will cause a level of damage deemed dangerous, the meteoroid threat can be mitigated.

For particles large enough to either penetrate a wall or cause a damaging plasma discharge, the meteoroid model relating the mass of a particle with the expected level of flux (Sec. 1.4) can be expressed as

$$\Phi = \Phi_0 m_p^{1-s} \tag{4.9}$$

The flux per unit area dictated by the minimum-sized particles that will damage a spacecraft is found by combining Eqs. (4.4), (4.7), and (4.9):

$$\Phi_p = G_\phi (v_{\mathrm{rel}}\cos\vartheta)^{-3(1-s)\gamma_p} \tag{4.10}$$

where $G_\phi$ is a constant, depending on the material properties of the wall and meteors, the flux constant $\Phi_0$, and whether physical penetration or plasma generation is being considered. The total number of damaging particles can then be summed over all the exposed portions of the vehicle ($A \cos \vartheta$ is the area exposed to the meteors for that panel $i$, $v_{\mathrm{rel}}$ is dominated by the meteoroid approach velocity $V_\infty$; $\Delta t$ and $V_\infty$ have been absorbed into $G_\phi$):

$$N_d = \Phi_p A_{\mathrm{total}} \Delta t = \sum_{i=1}^{\#\,\mathrm{panels}} G_{\phi i}(A_i \cos\vartheta_i)(\cos\vartheta_i)^{-3(1-s)\gamma_p} \tag{4.11}$$

### 4.2.2 Rectangular Spacecraft

Consider a box-like satellite with three distinct panels that can be exposed to the meteors (Fig. 4.1). It will be assumed that, for ease of the analysis, the spacecraft has the same type of shielding regardless of which panel is being considered. Therefore, $G_\phi$ will be identical for each area for a particular meteor event. For spacecraft with different types of shielding along different portions of the vehicle, this will not be the case, and the different material properties or thicknesses should be carried along in a more complete analysis.

Define a spacecraft-centered coordinate system with the $y$ axis pointed toward the meteors (note this should be the once-per-rev time-varying radiant in order to take full advantage of this mitigation option; Sec. 1.3). Because the problem is symmetric about this axis, it is immaterial where the $x$- and $z$-axis point. The only requirement is to constrain them to be perpendicular to the $y$ axis and to complete the standard right-hand system. Two rotation angles $\theta$ and $\phi$ completely define the orientation of the spacecraft with respect to the approaching particles. The angle $\theta$ is defined as a positive rotation about the $x$ axis in the $y$–$z$ plane (a pitch rotation). Angle $\phi$ is a positive rotation about the new $z'$ axis in the $x$–$y'$ plane (a yaw rotation). This completely describes a roll-pitch-yaw sequence because the coordinate system was chosen to be symmetrical about the roll axis. The projected areas exposed to the meteors become

$$\begin{aligned} A_1 \cos\vartheta_1 &= A_1 \sin\phi \cos\theta \\ A_2 \cos\vartheta_2 &= A_2 \cos\phi \cos\theta \\ A_3 \cos\vartheta_3 &= A_3 \sin\theta \end{aligned} \tag{4.12}$$

Substituting into Eq. (4.11) yields the number of damaging particles:

$$\begin{aligned} N_d = G_\phi[&A_1(\cos\vartheta_1)^{1-3(1-s)\gamma_p} + A_2(\cos\vartheta_2)^{1-3(1-s)\gamma_p} \\ &+ A_3(\cos\vartheta_3)^{1-3(1-s)\gamma_p}] \end{aligned} \tag{4.13}$$

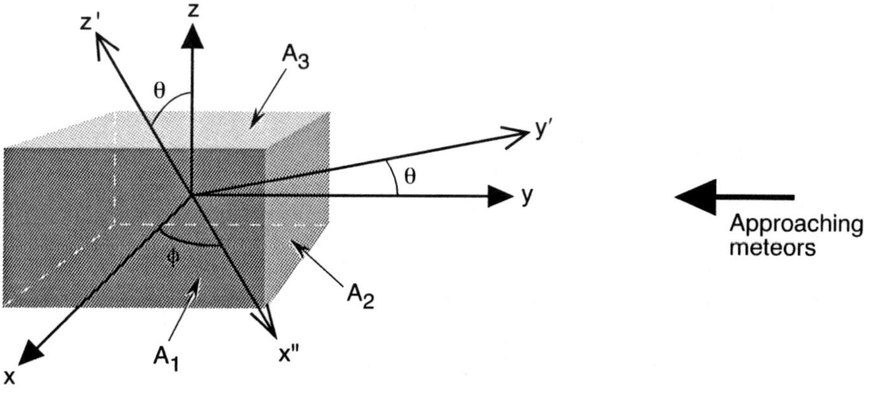


**Fig. 4.1. Box-like spacecraft coordinate system.**

When this expression is less than the damaging particles experienced by the minimal cross-sectional area when it is placed face-on to the meteors,

$$N_d < N_{d\min} = G_\phi A_{\min} \tag{4.14}$$

then canting the spacecraft away from the minimal cross-sectional area is more protective of the internal aspects of the vehicle. In essence, while more particles will hit a canted vehicle, there is less likelihood that they will produce spall or charge and thus cause damage (Fig. 4.2).

Because $G_\phi$ is assumed constant for each panel, the meteor properties and the material properties of the shielding will cancel, and the problem reduces to a geometrical one dependent upon the velocity index, $\gamma_p$, and the meteor stream mass index, $s$, (where $A_2$ is assumed to be smaller than both $A_1$ and $A_3$):

$$\begin{aligned}\frac{N_d}{G_\phi A_2} = {} & \frac{A_1}{A_2}(\sin\phi\cos\theta)^{1-3(1-s)\gamma_p} + (\cos\phi\cos\theta)^{1-3(1-s)\gamma_p} \\ & + \frac{A_3}{A_2}(\sin\theta)^{1-3(1-s)\gamma_p} < 1\end{aligned} \tag{4.15}$$

When this relation is satisfied, canting the spacecraft's surfaces provides greater protection than the minimal-area scenario in terms of particle penetration. The optimal orientation can be found by setting the derivatives with respect to $\theta$ and $\phi$ to zero and solving for the optimal angles:

$$\tan\phi_{\text{opt}} = \left(\frac{A_2}{A_1}\right)^{\frac{1}{-3(1-s)\gamma_p - 1}}$$

and (4.16)

$$\tan\theta_{\text{opt}} = \left(\frac{A_2}{A_3}\right)^{\frac{1}{-3(1-s)\gamma_p - 1}} \cos\phi_{\text{opt}}$$

The optimal angles are shown in Fig. 4.3 for $\gamma_p$ equal to two-thirds, $s$ is equal to 2.2 (i.e., Leonid outburst), and $A_1$ is assumed to be less than $A_3$. Note that while the optimal $\phi$ angle relies solely upon the single ratio of $A_2$ to $A_1$ (no $A_3$ dependence), the optimal $\theta$ angle relies upon two area ratios, owing to the dependence

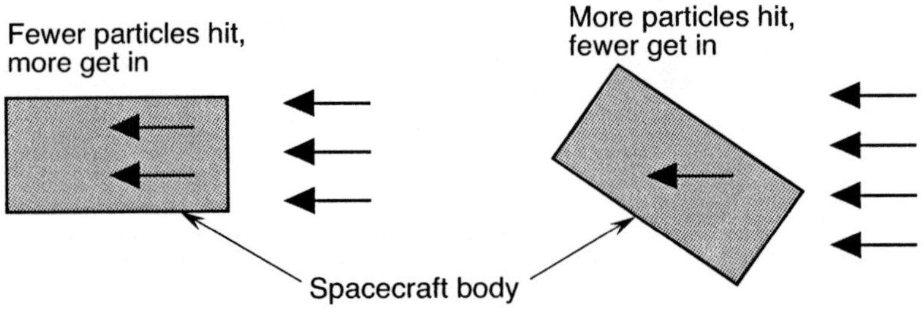


**Fig. 4.2. Penetration of shielding.**

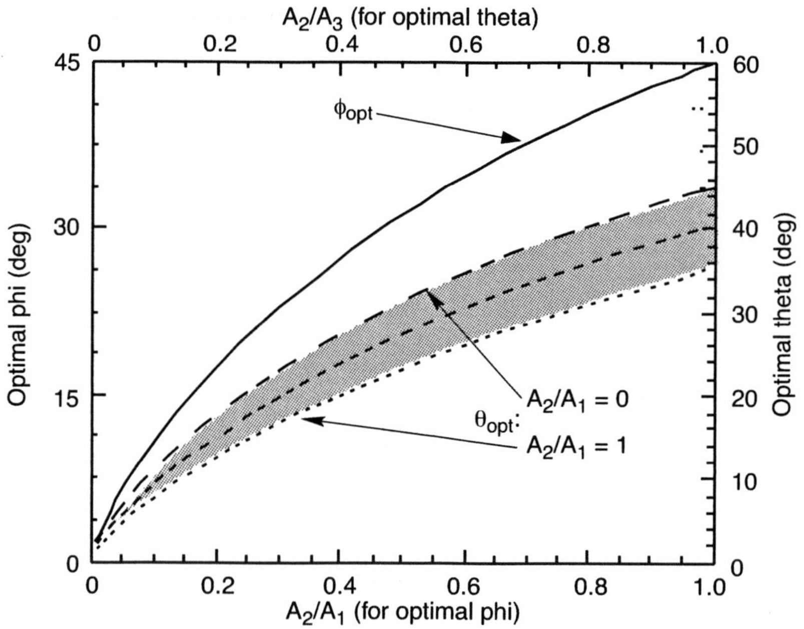


**Fig. 4.3. Optimal orientation angles for box.**

upon $\phi_{opt}$. As a consequence, the optimal $\theta$ is presented, having chosen a range of optimal $\phi$ angles ($A_2/A_1$ ranges from 0 to 1 in 0.5 increments, thus giving an envelope of possible values in the shaded region). It should be stressed that for every $\phi_{opt}$, there will be a corresponding curve for $\theta_{opt}$ somewhere in the envelope, but only a few are presented here for purposes of clarity. The bottom axis aligns with the left-hand axis ($\phi_{opt}$), and the top axis aligns with the right-hand axis ($\theta_{opt}$).

It should also be noted that $\gamma_p$ for penetration and $\gamma_p$ for plasma production will have different values for the same spacecraft. Therefore, the optimal orientation angles will also be different. This complicates planning, and a decision about trade-offs between the two phenomena is required.

For a cube-shaped satellite ($A_1 = A_2 = A_3$), $\phi_{opt}$ will be 45 deg, and $\theta_{opt}$ is 35.3 deg. This corresponds to orienting the cube such that the meteors are approaching along the diagonal. Counterintuitively, this implies that presenting the maximal cross-sectional area for a cubic spacecraft is actually more advantageous in terms of limiting the number of damaging particles that get past the shielding or create plasma than placing the minimal cross-sectional area flat on toward the storm. More particles will hit the diagonally oriented cube, but because the normal velocity is lower, fewer particles will get past the walls or create spall. Additionally, note that for a cube, the area ratios are unity, and therefore the optimal angles of Eq. (4.16) will not change regardless of the mass index $s$ or the velocity index $\gamma_p$. Thus, even though the ballistic limiting equations were experimentally derived for approach velocities of less than 10 to 12 km/sec and hence the actual equations for meteoric speeds are uncertain, the optimal angles for a cube will always be the same as presented here.

Although the optimal angles have been determined, it has not been shown that this orientation provides better protection than the minimal-area scenario. Substituting the optimal angles into the normalized number of damaging particles (Eq. [4.15]) yields

$$\left(\frac{N_d}{G_\phi A_2}\right)_{opt} = (\cos\phi_{opt}\cos\theta_{opt})^{-3(1-s)\gamma_p - 1} \qquad (4.17)$$

When this expression is less than unity, canting the spacecraft produces better performance in protecting the internal workings of the vehicle from meteoric damage. Because the cosine of any angle is always less than or equal to 1, Eq. (4.17) will be less than unity when the value of the exponent is greater than zero.

For presentation purposes, assume the same values for $A_2/A_1$ as depicted in Fig. 4.3 ($A_2/A_1 = 0.00, 0.50, 1.00$; $\gamma_p = 2/3$; $s = 2.2$). The optimal normalized number of damaging particles for these conditions is shown in Fig. 4.4 with the point for cubic spacecraft explicitly marked. Therefore, a cubic spacecraft oriented in this manner will suffer about 45% of the damage that it would otherwise experience if oriented with the minimal cross-sectional area facing the meteoroids.

But because all the values are less than unity, this implies that canting the vehicle will always provide better protection from penetration than placing the minimal cross-sectional area face on to the meteors regardless of vehicle shape (shaded area in Fig. 4.4). It must be noted that these figures assume that all of the spacecraft surfaces are of the same material and thickness, and protect portions of the spacecraft that have equal importance to the operator. If any of these constraints are violated, these generalized results will not apply. For example, the Hubble Space Telescope will orient itself during the 1998 and 1999 Leonids to

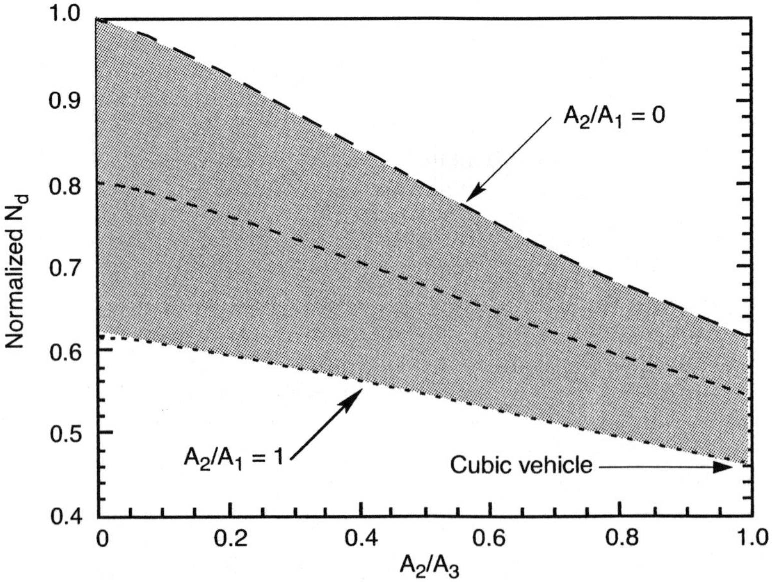


**Fig. 4.4. Normalized penetrating flux for optimal orientation.**

place the instrument package toward the meteors, effectively shielding the mirror with ~11 feet of buffer while only a few sheets of shielding protect the mirror from the side. The optimal $\theta$ in this case is less than a hundredth of a degree from the minimal cross-sectional area situation. Therefore, no appreciable benefit is gained by reorienting Hubble from the minimal attitude.

Earlier on, it was stated that there is a level of uncertainty in the value of the parameter $\gamma_p$ with a value of two-thirds chosen as typical for the presentation of the sample case given above. For this value of the velocity index, the optimal orientation for all area ratios provided better protection against penetration than the minimal cross-sectional area for the Leonids. But in a more general sense, the limiting value of the velocity index for any meteor stream can be solved for by setting the exponent of Eq. (4.17) to zero:

$$\gamma_{p-\mathrm{lim}} = \frac{1}{-3(1-s)} \tag{4.18}$$

When $\gamma_p$ is above this value, the optimal orientation provides better protection (the number of damaging particles of Eq. (4.17) is a minimum). When it is not above this limit, the optimal angles are worse than the minimal cross-sectional area (the number of damaging particles of Eq. (4.17) is a maximum) and therefore presenting the minimal cross-sectional area is better. The optimal angles will change, depending upon the value of $\gamma_p$, but the general principle will hold. With a minimum mass index typically thought to be around 2.2 (for the Leonid outbursts; Sec. 1.4), the largest value of the limiting velocity index would be 0.28. This is smaller than the index for typically used shields, although the multishock velocity index is close to one-third. It is also smaller than the velocity index typical for plasma production ($\gamma_p$~1.5). Thus, for the Leonids, the optimal orientation will yield greater protection than the minimal cross-sectional area. It should again be emphasized that this is contingent upon the form of the ballistic limiting equations and charge production equations (Eq. [4.4]) being valid for the Leonids. The actual behavior of the equations at ultrahigh velocities is uncertain at the current time.

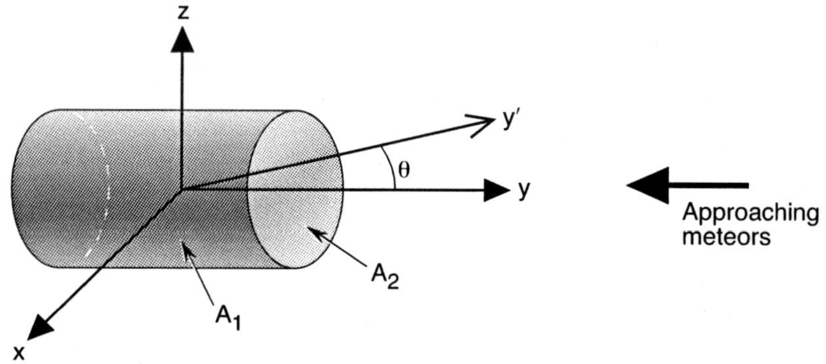


**Fig. 4.5. Cylindrical spacecraft coordinate system.**

### 4.2.3 Cylindrical Spacecraft

For a cylinder, the problem somewhat simplifies from the box case. Only two projected areas are relevant ($A_1$ and $A_2$, with $A_2$ being the minimal area facing the meteors) with a single relevant rotation angle $\theta$ because of symmetry about the approach vector (Fig. 4.5). The average normal velocity component for the curved area $A_1$ becomes

$$V_n = \frac{2}{\pi} V_\infty \sin\theta \tag{4.19}$$

Equation (4.15) describing the total normalized number of damaging particles for cylinders is

$$\frac{N_d}{G_\phi A_2} = (\cos\theta)^{1-3(1-s)\gamma_p} + \frac{A_1}{A_2}\left(\frac{2}{\pi}\sin\theta\right)^{1-3(1-s)\gamma_p} < 1 \tag{4.20}$$

which must be less than the total number of particles that would otherwise pierce the minimal area $A_2$ when that area is placed facing the meteors (i.e., the normalized number of particles must be less than unity). The optimal angle that provides the minimal number of damaging particles can be found in a similar manner to Eq. (4.16). Doing so for the cylindrical case yields

$$\tan\theta_{\text{opt}} = \left(\frac{A_2}{A_1}\right)^{\frac{1}{-3(1-s)\gamma_p - 1}} \cdot \left(\frac{\pi}{2}\right)^{\frac{1-3(1-s)\gamma_p}{-3(1-s)\gamma_p - 1}} \tag{4.21}$$

The corresponding normalized number of damaging particles is

$$\left(\frac{N_d}{G_\phi A_2}\right)_{\text{opt}} = (\cos\theta_{\text{opt}})^{-3(1-s)\gamma_p - 1} \tag{4.22}$$

Figure 4.6 shows the optimal angle as a function of the ratio of the areas for a mass index of 2.34 and velocity index of 2/3. Also shown is the normalized number of damaging particles at that angle, as well as the limiting value of $\gamma_p$. In comparing these curves with the box-like spacecraft (Figs. 4.3 and 4.4), it can be seen that the optimal $\theta$ angle of rotation is greater for the cylindrical case than the box with a corresponding greater measure of protection. Also, while the limiting value of $\gamma_p$ for the box was constant, the varying value indicated here shows that the limiting value of $\gamma_p$ is actually a function of the shape of the vehicle. However, the limit is still much lower than currently accepted bounds on $\gamma_p$, regardless of whether spall or ESD is being considered. As noted above, these results are good only for spacecraft that have similar walls all over the vehicle and where every portion of the protected components is equally valuable. For spacecraft with differing levels of protection and material properties in separate areas of the vehicle, a distinct analysis should be performed.

Therefore, depending upon the material properties and shape of the vehicle, it may actually be more advantageous to alter the orientation of the vehicle away from the minimal cross-sectional area position when trying to lessen the damage caused by an impact. With canted surfaces, the normal velocity component that drives the probability of penetration past a spacecraft's shields lessens, and so

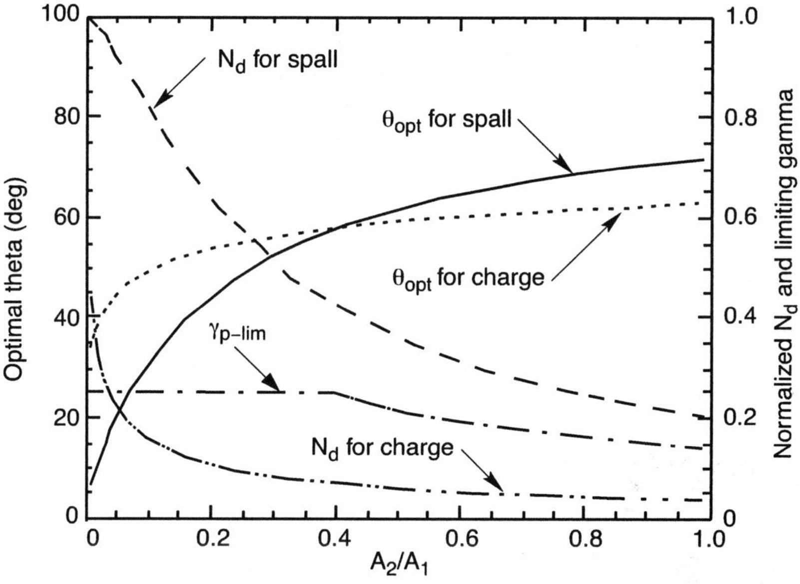


**Fig. 4.6. Optimal orientation angle and normalized number of damaging particles for cylinder.**

while a larger area is exposed, the actual number of particles that pierces the walls decreases. This study assumes that the walls of a vehicle are of the same material and thickness throughout the structure. If these conditions are not met, the analysis results will change. As a consequence, it must be emphasized that each specific type of spacecraft needs to be examined in order to determine the optimal orientation angles and the corresponding level of benefit gained for that spacecraft.

It should also be reiterated that the optimal angles for the minimization of plasma production for the same spacecraft will be different than for particle penetration. Figure 4.6 shows that the particle reduction is also much greater for the plasma production than for particle penetration. This is because of the larger values of $\gamma_p$ that exist for plasma (~1.5 for plasma versus ~2/3 for penetration). An approximate order of magnitude improvement could be achieved for the reduction of plasma-induced ESD.

It is useful to represent in a consistent manner the level of benefit gained from each of the strategies examined in this chapter. In the previous analysis, the benefit from this strategy was shown as a ratio between the number of damaging particles faced by the spacecraft when it was in the optimal orientation as compared with the minimal cross-sectional area. However, it will be more convenient for the other strategies to express the level of benefit as a reduction in the danger that is faced rather than as a strict ratio. Therefore, the level of benefit will be defined as the reduction in danger (either through damage or simple impact) referenced to the nominal condition. For this strategy, the level of benefit is unity minus the normalized number of damaging particles as previously computed through Eq. (4.17) and Eq. (4.22). For the rectangular spacecraft, the level of benefit is:

$$R_{DM} = 1 - \left(\frac{N_d}{G_\phi A_2}\right)_{opt} = 1 - (\cos\phi_{opt}\cos\theta_{opt})^{-3(1-s)\gamma_p - 1} \tag{4.23}$$

1. Optimal orientation angles for box:

$$\tan\phi_{opt} = \left(\frac{A_2}{A_1}\right)^{\frac{1}{-3(1-s)\gamma_p - 1}}$$

$$\tan\theta_{opt} = \left(\frac{A_2}{A_3}\right)^{\frac{1}{-3(1-s)\gamma_p - 1}} \cos\phi_{opt}$$

for cylinder: $$\tan\theta_{opt} = \left(\frac{A_2}{A_1}\right)^{\frac{1}{-3(1-s)\gamma_p - 1}} \cdot \left(\frac{\pi}{2}\right)^{\frac{1-3(1-s)\gamma_p}{-3(1-s)\gamma_p - 1}}$$

2. DeltaV requirements and mission downtime are minimal.
3. Benefit gained (amount of reduction in the penetrating flux for the optimal orientation compared with the minimal cross-sectional area attitude):

$$R_{DM} = 1 - \left(\frac{N_d}{G_\phi A_2}\right)_{opt} = 1 - (\cos\phi_{opt}\cos\theta_{opt})^{-3(1-s)\gamma_p - 1}$$

4. In general, when the velocity index is above a certain value dependent upon the mass index of the meteor stream, the optimal orientation provides more protection to the internal workings of the spacecraft than the minimal cross-sectional area:

$$\gamma_p > \frac{1}{-3(1-s)}$$

5. Applicability: all satellites with finite bodies, especially shielded vehicles.

**Algorithm 4.1. Attitude control for spall and ESD damage minimization.**

## 4.3 Attitude Control for Area Minimization

The number of particles that will impact a spacecraft during the peak of a meteor outburst is given by Eq. (1.44):

$$N = \Phi A_{s/c} \Delta t_{event} \tag{4.24}$$

where $N$ is the number of impacts, $\Phi$ is the flux, $A_{s/c}$ is the spacecraft area exposed to the meteors, and $\Delta t$ is the time of exposure. Therefore, the number of impacts can be reduced by lessening any one of these three quantities. This section examines reducing the exposed area through attitude orientation.

### 4.3.1 Vehicle Area Minimization

For vehicles with exposed sensors or equipment, the shield penetration of Sec. 4.2 may not be the dominant issue. Instead, it may be more important to minimize the number of particles that hit. In terms of minimizing the cross-sectional area, the vehicle body is the easiest factor to solve for and involves two options. The first is simply to point the minimal cross-sectional area toward the meteors in order to minimize the probability of dangerous impact. The second option does not necessarily minimize the presented area but instead points the most heavily shielded portion of the spacecraft at the approaching particles in order to protect the spacecraft's internal systems (e.g., the Hubble Space Telescope just examined). Because the right ascension and declination of the meteor storm radiant point is known from the orbital dynamics of encounter (Sec. 1.3), it is a simple matter to point the desired normal vector in the desired direction. It should be noted that for the best protection, the once-per-rev varying radiant of Eq. (1.22) should be used (repeated as Eq. [4.27]).

### 4.3.2 Solar Panels

The second spacecraft parameter that can be attitudinally controlled is the solar panels. The objective is to minimize the cross-sectional area of the panels with respect to the meteors, while at the same time maintaining as much solar power collection capability as possible. Instead of a simple minimization, the desire to maintain maximum photon absorption at the same time must also be considered.

Consider a spacecraft with a body of negligible size that has solar panels of a given area (i.e., with solar panels that are much bigger than the vehicle). The solar panels in this case are the main drivers in determining the danger of meteorite impact, while the probability of impact from the finite size of the nonpanel vehicle is hereafter ignored for the purposes of this section (i.e., negligible spacecraft cross-sectional area; vehicle attitude is assumed a separate issue from solar panel orientation; see the previous sections). The important parameters in the problem are the location of the radiant point of the incoming meteors (defined by the vector $L$), the unit vector from the spacecraft to the Sun (the vector $S$), and the orientation of the panels, which can be uniquely defined by the vector normal to the panels (vector $P$). All three vectors are shown in Fig. 4.7.

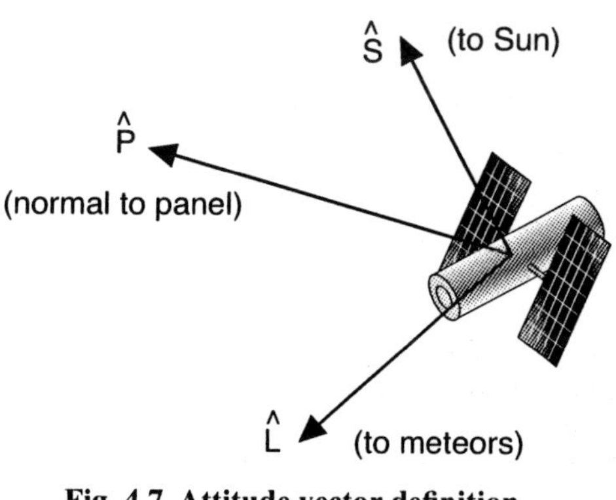


**Fig. 4.7. Attitude vector definition.**

The panel normal vector can be written in terms of the right ascension and declination as

$$\hat{P} = [\cos\delta_p \cos\alpha_p, \cos\delta_p \sin\alpha_p, \sin\delta_p]^T \tag{4.25}$$

where $\alpha_p$ and $\delta_p$ are the right ascension and declination of the panel vector measured in the geocentric equatorial coordinate frame. The direction of the incoming meteors is given in the same coordinate system as

$$\hat{L} = [\cos\delta_L \cos\alpha_L, \cos\delta_L \sin\alpha_L, \sin\delta_L]^T \tag{4.26}$$

where $\alpha_L$ and $\delta_L$ are the components of the time-varying meteor shower radiant given by Eq. (1.22), repeated here as Eq. (4.27):

$$\alpha_L = \tan^{-1}\left(\frac{v_{\mathrm{rel}\,y} - v_s \sin u \sin\Omega + v_s \cos u \cos\Omega \cos i}{v_{\mathrm{rel}\,x} - v_s \sin u \cos\Omega - v_s \cos u \sin\Omega \cos i}\right)$$

$$\delta_L = \sin^{-1}\left(\frac{v_{\mathrm{rel}\,z} + v_s \cos u \sin i}{[V_\infty^2 + 2(\bar{v}_{\mathrm{rel}} \cdot \bar{v}_s) + v_s^2]^{1/2}}\right) \tag{4.27}$$

where $u$ is the argument of latitude, $i$ is the inclination, and $\Omega$ is the right ascension of the ascending node, $v_s$ is the magnitude of the spacecraft velocity, and $v_{\mathrm{rel}}$ is the relative velocity between the Earth and the comet.

The unit vector to the Sun varies throughout the year but can still be written in terms of the right ascension and declination as

$$\hat{S} = [\cos\delta_s \cos\alpha_s, \cos\delta_s \sin\alpha_s, \sin\delta_s]^T \tag{4.28}$$

where both the right ascension and declination of the Sun move with a cycle of annual period.

The variation with the date occurs not only because of the radiant point motion but also because the Sun does not remain in a constant position and varies over the course of the year in a sinusoidal manner. This can be found to a high level of precision from the JPL ephemeris or approximately through analytical equations from the *Astronomical Almanac* as

$$\alpha_s = \lambda - \frac{180}{\pi}\tan^2\left(\frac{\varepsilon}{2}\right)\sin 2\lambda + \frac{90}{\pi}\tan^4\left(\frac{\varepsilon}{2}\right)\sin 4\lambda$$

$$\delta_s = \sin^{-1}(\sin\varepsilon \sin\lambda) \tag{4.29}$$

where $\lambda$ is the ecliptic longitude of the Sun given by

$$\lambda = 280°.461 + 0°.9856474n + 1°.915 \sin g + 0°.020 \sin 2g \tag{4.30}$$

ε is the obliquity:

$$\varepsilon = 23°.439 - 0°.000\,000\,4n \tag{4.31}$$

$g$ is the mean anomaly of the Sun:

$$g = 357°.528 + 0°.985\,600\,3n \tag{4.32}$$

and where $n$ is the number of days from J2000.0 (Julian date 2451545.0).

The angles between the vectors can be expressed in a manner similar to the usual beta and zeta angles used in eclipse studies (Chobotov, 1991; Chapter 2 of this document). The beta angle gives the angular separation between the orbit plane and the Sun vector, while the zeta angle in a similar manner describes the arc between the orbit plane and the vector from which the meteors are approaching (Fig. 4.8[a]). In the case under current study, the angles are not in reference to the orbit plane but rather the plane of the solar panels and are thus attitudinal in origin. For this reason, the two angles are hereafter referred to as the attitudinal beta and zeta angles ($\beta_A$ and $\zeta_A$, Fig. 4.8[b]). Note that the panel orientation problem being examined, although attitudinal in nature, is not transparent to the orbit elements of the vehicle because the meteor radiant moves as a function of the orbit elements. There will therefore be an implicit once-per-orbit variation in the optimal panel-pointing vectors.

The nondimensionalized, projected area with respect to each celestial phenomenon is simply computed as functions of the angular measurement of the respective attitudinal beta and zeta angles (because the area of the panels is constant, it can be divided out and the problem therefore nondimensionalized so as to be applicable to a solar panel of any given area):

$$\begin{aligned} J_1 &= \cos(90 - \zeta_A) = (\hat{L} \cdot \hat{P}) = \cos\delta_p \cos\delta_L \cos(\alpha_p - \alpha_L) + \sin\delta_p \sin\delta_L \\ J_2 &= \cos(90 - \beta_A) = (\hat{S} \cdot \hat{P}) = \cos\delta_p \cos\delta_s \cos(\alpha_p - \alpha_s) + \sin\delta_p \sin\delta_s \end{aligned} \tag{4.33}$$

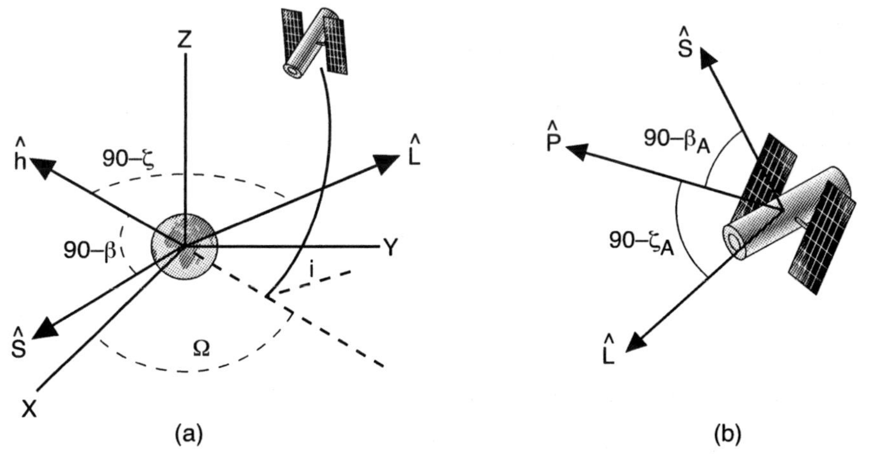


**Fig. 4.8. Beta and zeta angle definition for orbit plane and solar panels.**

The complete performance index must be constructed in such a way as to force $J_1$ to be equal to zero, while at the same time maximizing $J_2$. Such a scenario is desirable to make sure that the panel area as viewed by the meteors is nonexistent, while the Sun sees as much of the solar panel as possible. This is achieved by taking the sine of the attitudinal beta angle instead of the cosine. The cosine going to zero indicates that the vectors are perpendicular; but when the sun and panel vectors are coaligned, it is the sine that now goes to zero. In essence, maximize $J_2$ subject to the constraint $J_1 = 0$. The final performance index to be minimized is then

$$J = \lambda J_1 + (1 - J_2^2) \tag{4.34}$$

where $\lambda$ is a Lagrange multiplier. The multiplier, along with the optimal right ascension and declination of the panel, can be solved for by setting the derivatives of Eq. (4.34) with respect to those variables to be equal to zero:

$$\frac{\partial J}{\partial \lambda} = J_1 = 0 \quad \frac{\partial J}{\partial \delta_p} = 0 \quad \frac{\partial J}{\partial \alpha_p} = 0 \tag{4.35}$$

Solving gives the optimal pointing for the panel if it is desired to maximize the solar power collection while at the same time keeping the panel edge-on to the approaching meteors:

$$\begin{aligned} \tan\alpha_p &= \frac{\cos\delta_s \sin\alpha_s - \cos\delta_L \sin\alpha_L(\hat{L}\cdot\hat{S})}{\cos\delta_s \cos\alpha_s - \cos\delta_L \cos\alpha_L(\hat{L}\cdot\hat{S})} = \frac{y_s - y_L(\hat{L}\cdot\hat{S})}{x_s - x_L(\hat{L}\cdot\hat{S})} \\ \tan\delta_p &= -\frac{\cos(\alpha_p - \alpha_L)}{\tan\delta_L} \end{aligned} \tag{4.36}$$

where $x$ and $y$ are the unitized components of the Sun and meteor radiant vectors. Note the similarity in the right ascension expression with Eq. (4.27). The declination does not follow this pattern. The dot product of the meteor radiant and Sun vector is

$$(\hat{L}\cdot\hat{S}) = \cos\delta_p \cos\delta_L \cos(\alpha_p - \alpha_L) + \sin\delta_p \sin\delta_L \tag{4.37}$$

Note that the sense of the angles must be tested to make sure that the front of the solar panel points at the Sun. This occurs when the dot product between the Sun and panel vectors is greater than zero:

$$J_2 = (\hat{S}\cdot\hat{P}) > 0 \tag{4.38}$$

### 4.3.3 Application to the Leonids

For the Leonid meteor showers, the right ascension and declination are given for the entire shower as (Sec. 1.3)

$$\alpha_L = 145°.84 + 0°.424*\text{day}$$
$$\delta_L = 24°.67 - 0°.156*\text{day} \tag{4.39}$$

where *day* refers to the date in November (i.e., November 14), 0 hour would compute as 14.0, while hour 12 of November 17 would be 17.5 in Eq. (4.39).

However, the category of time of peak activity is what is of greatest interest. At that time, the Leonid radiant is $\alpha_L$ = 153.45 deg, $\delta_L$ = 21.87 deg without including the high-frequency, once-per-orbit dependency. At the same time, the solar radiant is $\alpha_s$ = 232.96 deg, $\delta_s$ = –19.09 deg. From Eq. (4.36), the optimal panel-pointing vectors can be represented as

$$(\alpha_p, \delta_p) = (235.07, -19.95) \tag{4.40}$$

This will result in a unitized area facing the Sun of over 99% while eliminating the danger from the meteors. However, one issue remains: the once-per-rev motion of the radiant that is the result of the satellite's orbit.

In examining the once-per-orbit motion, specific orbits must be chosen for presentation purposes. Assume, then, two sample orbits: one is an 800-km altitude, near-circular Sun-synchronous satellite ($a$ = 7178 km, $i$ = 98.60 deg, $\Omega$ = 156.93 deg), while the other is a near-equatorial geosynchronous satellite ($a$ = 42,240 km, $i$ = 5 deg). The once-per-orbit radiant motion variation in the Leonid

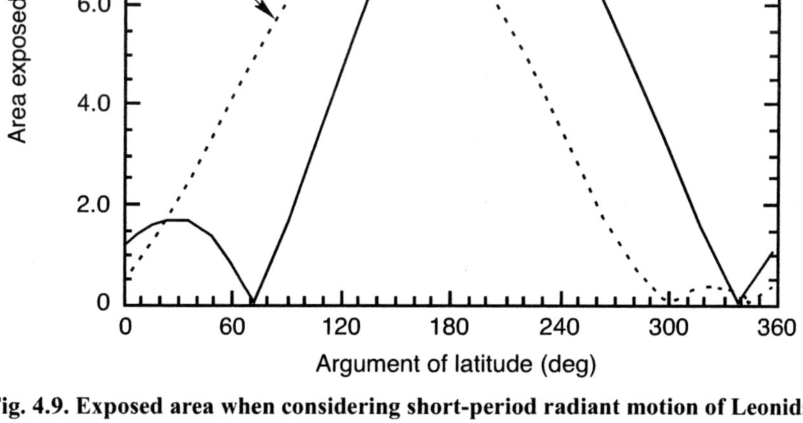


**Fig. 4.9. Exposed area when considering short-period radiant motion of Leonids.**

mean for this example was presented in Fig. 1.11. The general governing equation for the short-period radiant motion is given by Eq. (4.27). The variations in the radiant caused by the short-period effect can be on the order of several degrees in either component, depending upon the specific orbit geometry under consideration.

But what difference does the once-per-rev motion make in the area that is exposed to the meteors? Figure 4.9 shows the normalized area if the mean radiant is followed for the panel pointing instead of the once-per-rev varying radiant. That is, the meteors move in an apparent short-period sense, but the panels are pointed only considering the mean radiant of Eq. (4.40). Clearly, a significant level of exposure results when only the mean is used. If the solar direction is followed exactly without regard to the meteors, about 4% of the solar panel area will be exposed to the shower. The peak percentage shown here compares to about twice that level, depending upon the specific orbit. The once-per-orbit variation in the radiant point motion is therefore an important factor to consider when designing the attitude profile of the solar panels or any other spacecraft component. If it is desired to minimize the area that is exposed to the meteors, the short-period effects absolutely must be included in the optimal pointing determination. For the Leonid meteors, this short-period effect may be larger than the difference between the mean radiant and the solar direction.

Some of the most significantly dangerous meteor events are short in duration (e.g., 1 to 2 hours for the anticipated Leonid storms of 1998–2000). Some advantage can be taken from this in easing the complexity of the problem. Consider the near-equatorial GEO case separately from the LEO. For a 1- to 2-hour Leonid storm, an angle roughly about 30 deg in size will be traversed by a GEO satellite, as opposed to a complete orbit for a LEO satellite. If the argument of latitude for such a GEO satellite is located close to the zero points as depicted in Fig. 4.9, then following the simpler mean radiant will cause only about one-third of one percent of the solar panel area to be exposed to the particles. If the in-track location is not close to one of the zero points, a "mean" radiant adjusted for the once-per-rev motion but assumed constant over the storm interval can be followed. This would ease the complexity of the attitude control profile while still reducing the risk significantly. This would be applicable only for high-altitude, slowly moving vehicles, but does reduce the potential difficulty in planning the attitude profile of the spacecraft.

### 4.3.4 Sensors

The protection of vehicle sensors can be treated in several ways. The first method is simply to orient the spacecraft body such that the bulk of the vehicle is between the approaching meteors and the sensor. In this case, the vehicle shields the sensitive sensors from harm. A similar but potentially distinct strategy from this is to orient the sensors themselves to point away from the meteors so that, for example, the electronic package of a telescope shields the mirror. Instead of the vehicle providing the bulk of the protection, the sensor's equipment itself does the job. The Hubble Space Telescope uses a combination of these strategies and protects the optics by placing the back end of the spacecraft where the spacecraft systems reside toward the dominant meteor shower direction. In the first case, the pointing vector defining the bulk of the vehicle should be

1. Area presented to meteors: $A_L = \cos\delta_p \cos\delta_L \cos(\alpha_p - \alpha_L) + \sin\delta_p \sin\delta_L$

2. Area presented to Sun: $A_s = \cos\delta_p \cos\delta_s \cos(\alpha_p - \alpha_s) + \sin\delta_p \sin\delta_s$

3. Optimal panel pointing angles for meteor area to be zero while also maximizing the solar area:

$$\tan\alpha_p = \frac{\cos\delta_s \sin\alpha_s - \cos\delta_L \sin\alpha_L(\hat{L}\cdot\hat{S})}{\cos\delta_s \cos\alpha_s - \cos\delta_L \cos\alpha_L(\hat{L}\cdot\hat{S})} = \frac{y_s - y_L(\hat{L}\cdot\hat{S})}{x_s - x_L(\hat{L}\cdot\hat{S})}$$

$$\tan\delta_p = -\frac{\cos(\alpha_p - \alpha_L)}{\tan\delta_L}$$

4. DeltaV requirements and mission downtime are minimal.

5. Benefit gained will of course be 100% if the meteor area is zero; if it is not, then the benefit gained through any specific pointing $(\alpha_p, \delta_p)$ is (where $A_L(S)$ is the meteor projected area when the solar panels are pointing directly at the Sun, the flux and the time of exposure are constant, and $N_0$ is the nominal number of impacts):

$$R_{AM} = \frac{N_0 - N_{\text{reduced}}}{N_0} = 1 - \frac{A_L}{A_L(S)}$$

6. Applicability: all satellites with solar panels.

**Algorithm 4.2. Solar panel area minimization.**

pointed at the meteor radiant (Eq. [4.27]); in the second, the sensor should be pointed toward the meteor antiradiant (opposite of Eq. [4.27]).

Another option is to orient the sensors such that a minimal cross-sectional area is presented to the meteors. If the sensor is fairly flat (as is an antenna), the minimal area will be when the sensor-pointing vector is perpendicular to the meteor radiant vector. But the most interesting option is to try to maintain mission operational status, while at the same time minimizing the cross-sectional area presented to the meteors. If a sensor type is assumed that is large and fairly flat (e.g., an antenna), it is a simple matter to substitute the desired pointing vector into Eq. (4.36) instead of into the Sun-pointing vector ($\alpha_s$, $\delta_s$) and generate the corresponding optimal pointing vector ($\alpha_p$, $\delta_p$).

For an Earth-observing satellite in a near-circular orbit, the desired pointing vector goes from the satellite to the Earth opposite in direction to the position vector of the vehicle. Thus, a relation is established between the right ascension and declination of the desired attitude and the orbit elements of the spacecraft ($\alpha_s$ and $\delta_s$ are still used here to be notationally consistent with earlier equations, but should not be confused with the Sun; instead, $\alpha_s$ and $\delta_s$ here refer to the direction from the vehicle to the center of the Earth):

$$\hat{r}_s = -\begin{bmatrix} \cos\alpha_s\cos\delta_s \\ \sin\alpha_s\cos\delta_s \\ \sin\delta_s \end{bmatrix} = \begin{bmatrix} \cos u\cos\Omega - \sin u\sin\Omega\cos i \\ \cos u\sin\Omega + \sin u\cos\Omega\cos i \\ \sin u\sin i \end{bmatrix} \tag{4.41}$$

Knowing the inclination ($i$) and the node ($\Omega$) at the time of the storm for a particular vehicle, substitution of Eq. (4.41) into Eq. (4.36) will establish an optimal attitude profile as a function of the argument of latitude $u$. Note that the short-period dependence will exist not only because of the desired point angles ($\alpha_s$, $\delta_s$) but also because of the once-per-rev radiant point variation ($\alpha_L$, $\delta_L$), although this short-period change in the radiant will be small in comparison with Eq. (4.41).

However, for purposes of this generalized study, the node and inclination are not fixed, so generalized solutions to this problem are difficult to represent. As an example, however, assume a geosynchronous satellite in a near-circular, near-equatorial orbit ($i = 5$ deg in the example shown below). In this case, the small angle approximation can be used for the inclination. Because the desired sensor declination will never be larger than the inclination, the small angle approximation can also be used for $\delta_s$. An additional substitution can be utilized for the node and argument of latitude in terms of the true longitude $\ell$:

$$\ell = \omega + f + \Omega = u + \Omega \tag{4.42}$$

Equation (4.41) then yields

$$-\begin{bmatrix} \cos\alpha_s \\ \sin\alpha_s \\ \delta_s \end{bmatrix} = \begin{bmatrix} \cos\ell \\ \sin\ell \\ i\sin\ell \end{bmatrix} \tag{4.43}$$

Assume that the orbital period of interest takes place at the time of the storm (Leonid radiant is time varying in the short-period sense). Substituting into Eq. (4.36), numerically solving for one orbital period ($\ell$ varies through one cycle) will yield the optimal sensor vector if Earth pointing is desired (Fig. 4.10). This vector is the closest that the sensor can point to the Earth, while at the same time remaining perpendicular to the oncoming meteors. Also shown for comparison purposes in Fig. 4.10 is the vector pointing from the satellite to the Earth's center. Recall that this is the pointing vector desired for mission operations. It should be noted that with GEO satellites and a 2-hour Leonid storm with a similar uncertainty, the satellite will not move through a complete orbital cycle but rather will go through an arc of 30 deg ± 30 deg. The satellite will therefore cover only a portion of Fig. 4.10 during a short-duration storm.

The maximum angular difference between pointing at the Earth and in the optimal direction is about 45 deg. Therefore, maintaining operations status through the full length of the event is not practical, but this is dependent upon the position of the satellite in its orbit. Because the storm is of short duration and GEO satellites move slowly, a GEO satellite will again not go through a complete orbit but rather will only move through a 30-deg arc. If the in-plane location is in the proper place at $\alpha_L \pm 90$ deg, continuing the spacecraft's mission may be possible.

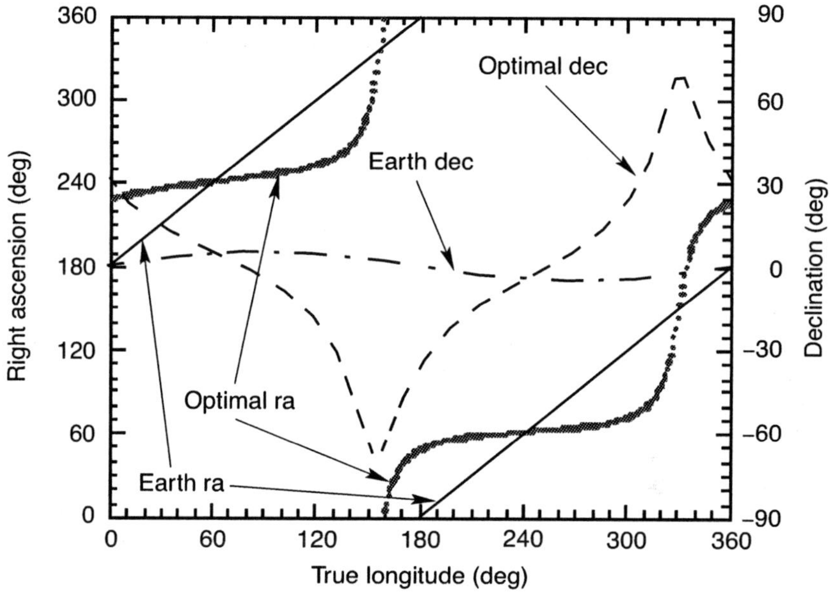


**Fig. 4.10. Optimal sensor right ascension and declination.**

## 4.4 Flux Minimization

The number of particles that will impact a spacecraft during the peak of a meteor outburst is given by Eq. (4.24) and repeated here as Eq. (4.44):

$$N_I = \Phi A_{s/c} \Delta t_{\text{event}} \tag{4.44}$$

To minimize this number, the spacecraft area can be reduced (Sec. 4.3) and the time spent in the event can be lessened through Earth shadowing (Chapter 2 and Sec. 4.6). However, another possibility exists in lowering the flux value a spacecraft experiences through its orbital location. If the vehicle is on the side of the Earth away from the core of the meteor stream or one of the smaller streamlets, then it presumably would be safer than if it were on the side toward the core (Fig. 4.11).

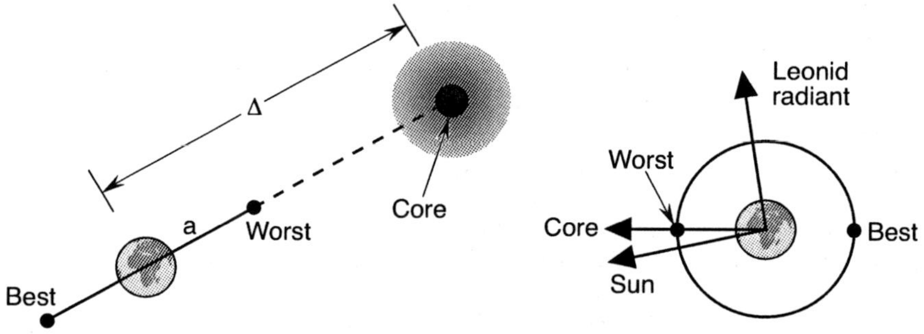


**Fig. 4.11. Earth-satellite-stream core geometry.**

Assume that the flux follows the same type of spatial distribution as its temporal resolution (i.e., a form similar to the ZHR activity curve of Eq. [1.25]):

$$\Phi = \Phi_c 10^{-D\Delta} \tag{4.45}$$

where $\Phi_C$ is the flux at the core, $\Delta$ is the distance from the core, and $D$ is the exponent computed from the radial component of the activity curve. Here, $D$ can be approximated by assuming that the difference between subsequent outbursts driven by the same phenomenon (e.g., near-comet 1966–7 and 1998–2000 Leonid outbursts) is the result of the difference in the close-approach distances of the two occurrences. This is only a rough approximation, because other factors are also significant (Sec. 1.4), but it does provide a starting point for the analysis. For the 1966 Leonid events, the ZHR has been estimated at anywhere from 15,000 to 150,000. Choose, then, the two most extreme values and assume the reality is somewhere in between. The 1998–2000 worst-case Leonids are expected to have a ZHR of approximately 5000. The $D$ values for the two extreme examples of 1966 storms coupled with the expected 1998–2000 storm are shown in Table 4.1. Recent computations of the 1999 Leonid encounter show that the streamlet generated in 1899 will cause the greatest problems (Brown *et al.*, 1999) and will be located to the outside of the Earth's orbit. In this case, the flux estimates will yield a value of $D$ of ~ 4500 (Cooke and Suggs, 1999).

Also shown is the relative benefit gained from a GEO satellite on the side of the orbit farthest from the core rather than closest. GEO satellites, with the largest semimajor axis, will gain the most from this strategy. Conversely, a LEO satellite will gain the least. Specifically, this is the amount of reduction in the flux that a spacecraft would experience by following this strategy. The benefit is computed:

$$\frac{\Phi_1 - \Phi_2}{\Phi_1} = 1 - 10^{-2\mathrm{Da}\cos\varphi} \tag{4.46}$$

where $\Phi_1$ and $\Phi_2$ are the fluxes at the safest and least safe points and where $a$ is the semimajor axis in the same units as $D$. This expression assumes that the cross-sectional area, $A$, and the time of exposure, $\Delta t$, are constant. Figure 4.11 shows the

**Table 4.1. Benefits of Flux Control for Collinear Geometry**

| Event | Max ZHR | Closest-approach distance (AU) | D (/AU) | LEO collinear benefit (%) | GEO collinear benefit (%) | GEO equatorial benefit (%) |
|---|---|---|---|---|---|---|
| 1998 | 5000 | 0.0085 | – | – | – | – |
| 1966a | 15,000 | 0.0035 | 95 | 2 | 12 | 9 |
| 1966b | 150,000 | 0.0035 | 296 | 7 | 32 | 26 |
| 1999 | 1000 | 0.0008 | ~4500 | 67 | ~99 | ~99 |

orbits as being coplanar with the direction of the core at closest approach, and thus represents the best possible values. Actual cases will reflect values that are somewhat lower. When the orbit is not coplanar with the Earth-core vector, an adjustment must be allowed to account for the fact that the maximum distance will not be the semimajor axis. This is represented by the cos φ factor in the exponent. It is found from the vector lying in the orbit plane that comes closest to the Earth-core vector dotted into the Earth-core vector itself:

$$\cos\varphi = [(\hat{h} \times \hat{R}_{ec}) \times \hat{h}] \cdot \hat{R}_{ec} \tag{4.47}$$

where $R_{ec}$ is the unitized vector from the Earth to the core at closest approach found from the ephemeris generated in Sec. 1.1 (equals [–0.5524, –0.6787, –0.4839] for the Leonid 1998 encounter in geocentric equatorial [ECI] coordinates), and $h$ is the unitized angular momentum vector of the orbit (Eq. [2.2]).

Assuming a 1998 encounter geometry and a near-equatorial GEO, the benefits are reduced to 9% and 26% for the two extreme values of the 1966 storm. Still, for heavy outbursts, this much of a gain could be of substantial importance. In 1999, the streamlet generated in 1889 will pass very close to, but just outside of, the Earth's orbit. The benefit thus gained can be extremely large for this case (>90%). Note that the level of benefit, because it is expressed as a percentage in the reduction of the flux, is independent of what the actual flux is. It merely represents the relative benefit gained by being on one side of the orbit versus the other.

It must be noted that the effectiveness of this strategy will be either enhanced or diminished in concert with the focusing effect of Chapter 3. However, for the Leonids and a near-equatorial GEO, the increase in the flux will only maximize, not in the direction closest to being opposite to the core, but in the direction closest to the opposite of the radiant vector. The relative benefits and disadvantages are therefore not directly additive. Even so, from Eq. (3.37), a near-equatorial GEO will experience a maximum focusing of the flux only to the 4% level, well below the expected benefits presented here for the core-distance effect.

The two in-plane locations of the closest and furthest points can be found easiest by equating the $z$ components of the satellite position vector (Eq. [4.41]) with the $z$ component of the vector in the orbit plane that comes closest to lining up with the Earth-core vector (Eq. [4.47]):

$$\sin u \sin i = [(\hat{h} \times \hat{R}_{ec}) \times \hat{h}]_z \tag{4.48}$$

yielding (making sure the sense of the angle is correct, i.e., the optimal desired is actually the minimum flux and not the maximum):

$$\sin u_{\text{opt}} = \sin i R_z + \cos i[\cos\Omega R_y - \sin\Omega R_x] \tag{4.49}$$

where $R_x$, $R_y$, and $R_z$ are the unitized components of the Earth-core vector at closest approach. For the near-equatorial GEO, the optimal in-plane location from Eq. (4.48) will naturally be at $(-R_x, -R_y, 0)$. For the 1998 Leonids, this corresponds to a true longitude of 50 deg E. This will change for future encounters depending upon the core's or streamlet's location.

In general, and in order to be consistent with the benefit ratios from the other sections, the relative benefit during the course of an orbital cycle is better referenced to the flux at the Earth $\Phi_0$. The benefit then becomes:

$$\frac{\Phi_0 - \Phi}{\Phi_0} = 1 - 10^{-\mathrm{Da}\cos\varphi} \tag{4.50}$$

where $\varphi$ is now the angle between the radius vector at any point in the orbit (Eq. [4.41]) and the Earth-core vector:

$$\cos\varphi = \hat{r}_s \cdot \hat{R}_{ec} \tag{4.51}$$

The deltaV requirements for in-plane maneuvering that might be necessary are discussed more fully in Sec. 4.5. In the extreme case where out-of-plane maneuvering is desired, these deltaV requirements are discussed in Sec. 4.6.

1. Compute stream flux exponent from historical data or from the radial component of the activity curve: $\Phi = \Phi_0 10^{-D\Delta}$
2. Optimal in-plane location: $\sin u_{\text{opt}} = \sin i R_z + \cos i[\cos\Omega R_y - \sin\Omega R_x]$
3. DeltaV requirements and mission downtime are related to the change needed to the satellite's in-plane location ($\Delta u$ is in degrees, $\Delta t$ and $TP$ are in days, velocities in km/s; this expression assumes a circular orbit with period $TP$):

$$\Delta v_{\text{total}} = \frac{8}{3}\frac{TP}{\Delta t_{\text{downtime}}}\frac{\Delta u}{360}v_s$$

4. Benefit gained (where the area and exposure time are constant):

$$R_{\text{FM}} = \frac{N_0 - N_{\text{reduced}}}{N_0} = 1 - 10^{-\mathrm{Da}\cos\varphi}$$

where $\cos\varphi = \hat{r}_s \cdot \hat{R}_{ec}$

5. Applicability: GEO benefits most, followed by HEO and MEO. LEO satellites can benefit, but not by much. Mission operation requirements may limit this strategy to on-orbit spares.

**Algorithm 4.3. Orbit maneuvering for flux minimization.**

## 4.5 Relative Velocity Minimization

The ballistic limiting equations of Sec. 4.2.1 show that the damage done to spacecraft from meteoroids is driven by the magnitude of the velocity normal to the surface. This is true regardless of the component being examined—vehicle, solar panel, or sensor. That section examined attitude control to lessen the normal component without changing the velocity magnitude. But a similar result can be found in another manner—by moving the spacecraft such that its velocity motion is traveling in the same direction as the meteor particles, thereby lessening the magnitude of the relative velocity at impact (Fig. 4.12, where the day-night terminator is included for the Leonids).

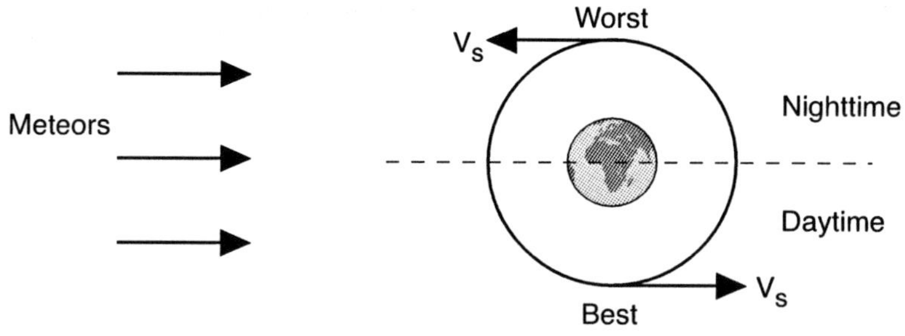


**Fig. 4.12. In-plane geometry for relative velocity minimization.**

Although the specific case of the highly eccentric Molniya orbits will be briefly discussed, this section will focus primarily upon MEO- and GEO-type orbits. LEO orbits are not viable for this strategy; they are so low that when comparing their orbital periods with the expected length of the storm (1 to 2 hours) with a similar length of time attributed to the uncertainty of the storm's arrival, the velocity vectors of the LEO satellites will go through an approximate orbit cycle. Therefore, their relative motion cannot be controlled within the error bars of the storm to the extent that it makes their use feasible. For this strategy to work, a fairly high-altitude satellite is required that has a nearly constant velocity vector over the time interval of the storm and its associated uncertainties.

For a near-circular orbit, the velocity can be written in Cartesian geocentric coordinates in terms of the orbital elements as:

$$\bar{v}_s = v_s \begin{bmatrix} -\sin u \cos\Omega - \cos u \sin\Omega \cos i \\ -\sin u \sin\Omega + \cos u \cos\Omega \cos i \\ \cos u \sin i \end{bmatrix} \tag{4.52}$$

where $v_s$ is the circular velocity of the satellite ($a$ is the semimajor axis, $\mu$ the gravitational parameter for Earth):

$$v_s = \sqrt{\frac{\mu}{a}} \tag{4.53}$$

and where $\Omega$, $i$, and $u$ are the right ascension of ascending node, inclination, and argument of latitude.

Because the meteor particles approach the Earth from a radiant point with a known right ascension and declination ($\alpha$ and $\delta$ are 153.45 deg and 21.87 deg for the Leonids) with known approach velocity ($V_\infty$= ~70 km/sec for the Leonids), the velocity of the particles in the Cartesian geocentric system is:

$$\bar{V}_\infty = -V_\infty \begin{bmatrix} \cos\alpha\cos\delta \\ \sin\alpha\cos\delta \\ \sin\delta \end{bmatrix} \tag{4.54}$$

The magnitude of the relative velocity between the meteors and the spacecraft becomes:

$$v_{\mathrm{rel}}^2 = (\bar{v}_s - \bar{V}_\infty)\cdot(\bar{v}_s - \bar{V}_\infty) \tag{4.55}$$

In order to minimize the damage from a collision, the relative velocity magnitude must be minimized. The ballistic limiting and ESD equations of Sec. 4.2.1 describing the damage are actually written in terms of the relative velocity magnitude raised to the power $\gamma_p$. However, irrespective of the value of $\gamma_p$, minimizing the relative velocity magnitude minimizes the damage the spacecraft will face.

Since both the spacecraft in a near-circular orbit and the particles approaching from a consistent position in the sky can be assumed to have constant speeds, the function to be minimized is the dot product of the two velocity vectors in terms of the orbit elements:

$$\frac{\partial(\bar{v}_s\cdot\bar{V}_\infty)}{\partial u} = \frac{\partial(\bar{v}_s\cdot\bar{V}_\infty)}{\partial\Omega} = 0 \tag{4.56}$$

Evaluating the normalized dot product yields:

$$\frac{(\bar{v}_s\cdot\bar{V}_\infty)}{v_s V_\infty} = \sin u\cos\delta\cos(\Omega-\alpha) - \cos u\sin\delta\sin i + \cos u\cos\delta\sin(\Omega-\alpha)\cos i \tag{4.57}$$

Solving Eq. (4.56) using Eq. (4.57) gives the expressions for the optimal node and argument of latitude as:

$$\sin(\Omega-\alpha) = -\frac{\tan\delta}{\tan i} \tag{4.58}$$

$$\tan u = \frac{\cos\delta\cos(\Omega-\alpha)}{\cos\delta\sin(\Omega-\alpha)\cos i - \sin\delta\sin i} \tag{4.59}$$

Equation (4.58) implies that the optimal plane of the orbit lies edge-on to the oncoming meteors, or in other words, that the orbit plane contains the meteor approach vector. However, it is not likely that satellite operators for higher-altitude satellites are going to be moving the node in any significant sense (Sec. 4.6 will examine this in detail). Therefore, Eq. (4.58) is inappropriate for this analysis and can be ignored (other than noting its physical significance as the optimal node location). As a consequence, the only orbital parameter available for deliberate change is the in-plane angle, the argument of latitude $u$.

Equation (4.59) for the argument of latitude admits two solutions—the minimum (lowest relative velocity) and the maximum (largest relative velocity). These two solutions will be 180 deg apart. A singularity exists when the denominator goes to zero; physically, this corresponds to the plane of the orbit being perpendicular to the meteor radiant. In this case, no minima or maxima will exist. Figure 4.13 shows the argument of latitude as a function of the node and the inclination for the minimum solution, and gives the locations where the velocity vector of the satellite comes closest to lining up with the meteor particle velocity vector and is headed away from the storm. Note the singularity where the contours come together; this represents the orbit plane being perpendicular to the meteors where no optimal solution exists. Again, the maxima (worst location) will be 180 deg away from the values shown here.

### 4.5.1 Effectiveness

Figure 4.13, while showing the optimal argument of latitude, does not yield information as to how effective it would be to have the satellite in the best configuration (headed away from the storm) versus being in the worst location (headed into the storm). Because the velocity of the meteors is much larger than any Earth-orbiting vehicle (~70 km/sec), the relative velocity magnitude can be normalized with respect to the orbital average value, and the variation about the orbit can subsequently be treated as an epsilon difference from the average. Since it can be assumed that the orbital velocity for near-circular orbits is constant, the $v_s^2$ terms in Eq. (4.55) can also be included in the normalization. The relative velocity magnitude equation then becomes:

$$v_{\text{rel}} = \left[(v_s^2 + V_\infty^2)\left(1 - 2\frac{\bar{v}_s \cdot \bar{V}_\infty}{v_s^2 + V_\infty^2}\right)\right]^{1/2} \tag{4.60}$$

The expression driving the orbital variation in the relative velocity magnitude, then, is simply the dot product of the spacecraft and meteor velocity vectors. Owing to symmetry, the maximum dot product of the vectors will be equal and opposite to the minimum dot product. Additionally, since the spacecraft velocity for the minimum relative velocity solution must be headed away from the approaching meteors, the dot product for the minimum solution will always be positive. Substituting into Eq. (4.10), the ratio between the minimum and maximum number of damaging particles over the course of an orbit becomes:

$$R_{\text{RV}} = 1 - \frac{N_{d-\min}}{N_{d-\max}} = 4\left[-\frac{3}{2}(1-s)\gamma_p\right]\left(\frac{\bar{v}_s \cdot \bar{V}_\infty}{V_\infty^2}\right) \tag{4.61}$$

Equation (4.61) thus gives a measure of the relative benefit that a vehicle gains by moving with the particles as opposed to moving into the meteor stream. With a mass index dependence as well as a velocity index dependence, it is difficult to represent Eq. (4.61) in a general manner. However, consider setting the expression in the brackets equal to unity. The physical significance of choosing the expression in this manner is apparent from Eq. (4.10). When the bracket is unity, the velocity index dependence of Eq. (4.10) goes to the second power of the velocity magnitude; the relative benefit ratio for the minimum and maximum number of damaging particles in this situation is therefore equivalent to the benefit ratio for the minimum and maximum kinetic energies. For the Leonid meteors with an approximate mass index of 2.2, the velocity index would be 0.56. This is within the typical ranges for the velocity index (Sec. 4.2.1). This is a specific case but is useful for general representation purposes, and for relating the

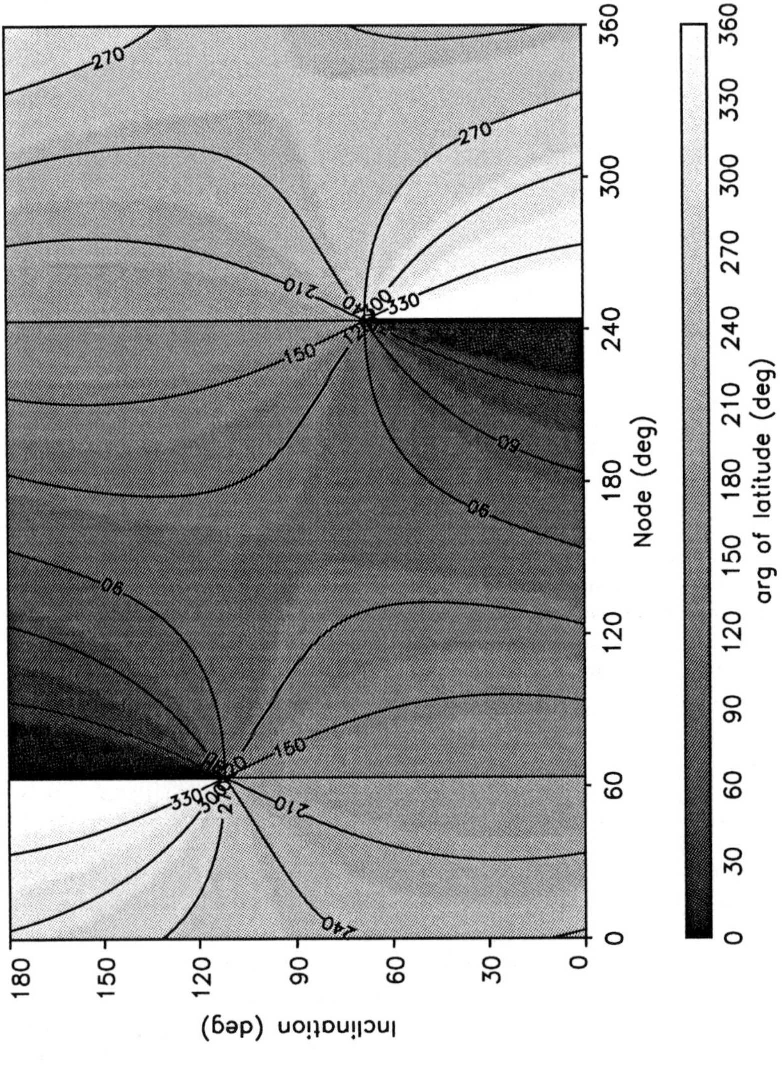


**Fig. 4.13. Minimal (lowest relative velocity) argument of latitude locations.**

results as a function of the kinetic energy at impact is easily understandable from a conceptual viewpoint. Assuming the spacecraft velocity is small with respect to the meteoroid velocity gives the kinetic energy ratio (Eq. [4.62]):

$$R_{\mathrm{KE}} = \frac{\mathrm{KE}_{\mathrm{min}}}{\mathrm{KE}_{\mathrm{max}}} = 4\left(\frac{\bar{v}_s \cdot \bar{V}_\infty}{V_\infty^2}\right) \tag{4.62}$$

Figures 4.14 through 4.16 show the kinetic energy ratio for LEO, MEO, and GEO orbits, respectively. LEO orbits are not of concern in this type of analysis because the storm is likely to be on the order of an approximately 2-hour duration, and therefore orbital periods that are a significant fraction of this (or even larger than the storm duration) will cycle through a complete orbit with the result that any in-track benefits (or detriments) will average out to a large degree. In essence, it is desirable to have the satellite moving slowly enough that the velocity vector stays predominantly in the same direction over the duration of the meteoroid storm. For this reason, this analysis is useful mostly for MEO and GEO orbits. Still, a representative example of a LEO orbit is shown here for comparison purposes.

For GEO orbits, the kinetic energy at impact can be reduced by ~18% between the worst in-plane location as compared to the best. For near-equatorial GEO orbits ($i < 11$ deg), the ratio ranges in a near-constant manner from 14% to 17%, regardless of the node, implying that a significant decrease in the kinetic energy at impact can be found through in-plane location of the satellite without concern for the nodal location.

For MEO orbits, the ratio grows to a maximum of 24%. If the GPS inclination of 55 deg is examined more closely, the ratio ranges between 5% and 22%, depending upon the node location. However, this strategy has limited applications for a constellation such as GPS. If the integrity of the constellation is to be maintained, the four satellites in each of the six planes must be spaced evenly. As such, any benefits from a lessened impact energy are canceled to a large degree by the satellite on the opposite side of the orbit, which will experience a greater impact energy with the particles (discussed in detail later on in Sec. 4.5.6). Therefore, the utilization of this strategy for any large constellation will cause a disruption of the entire configuration, most likely making this method impractical except in extreme circumstances where loss of vehicle would otherwise be very likely to occur. Note the similarities between these figures and Figs. 2.23–2.26. The best nodal locations for shadowing are also the best locations for relative velocity minimization.

For LEO satellites, the kinetic energy ratio achieves its largest range of values. While this strategy is impractical for LEO spacecraft, it is interesting to note that the maximum kinetic energy ratio depends solely upon the altitude. The maximum values of the kinetic energy ratio occur when the velocity vectors of the satellite and the meteors line up. This reduces Eq. (4.62) to a function of the semimajor axis only and is plotted in Fig. 4.17.

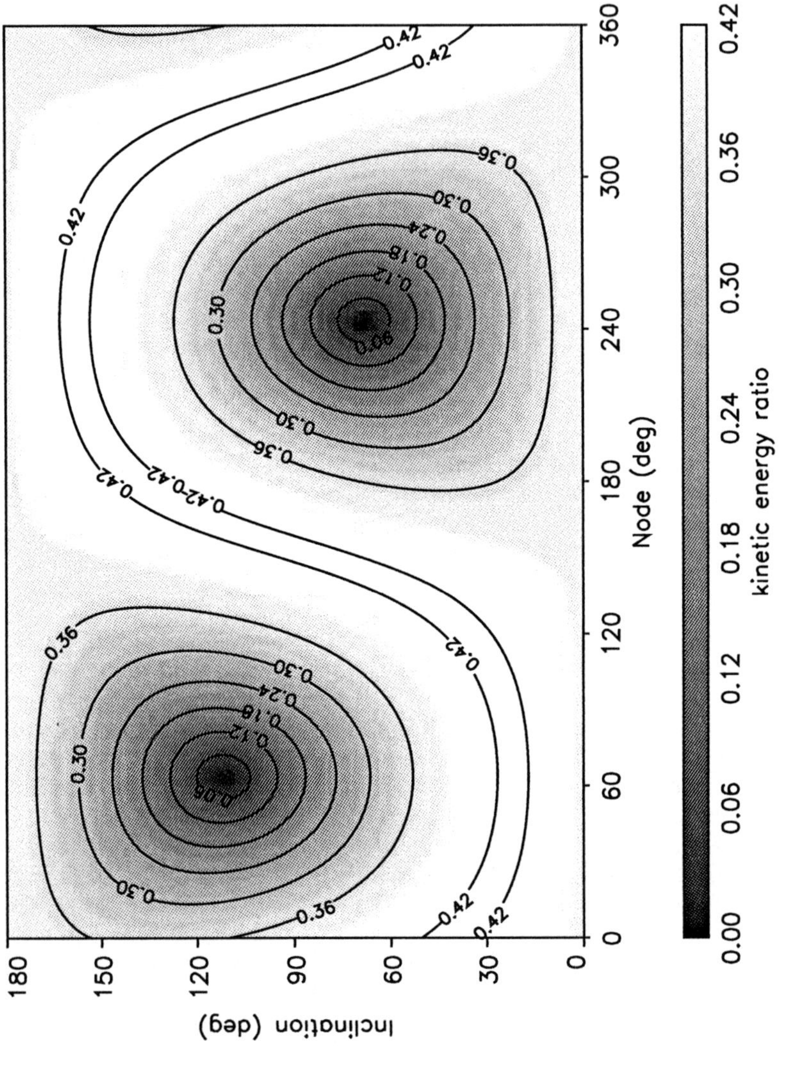


**Fig. 4.14. Kinetic energy ratio for LEO altitude orbits (altitude = 800 km).**

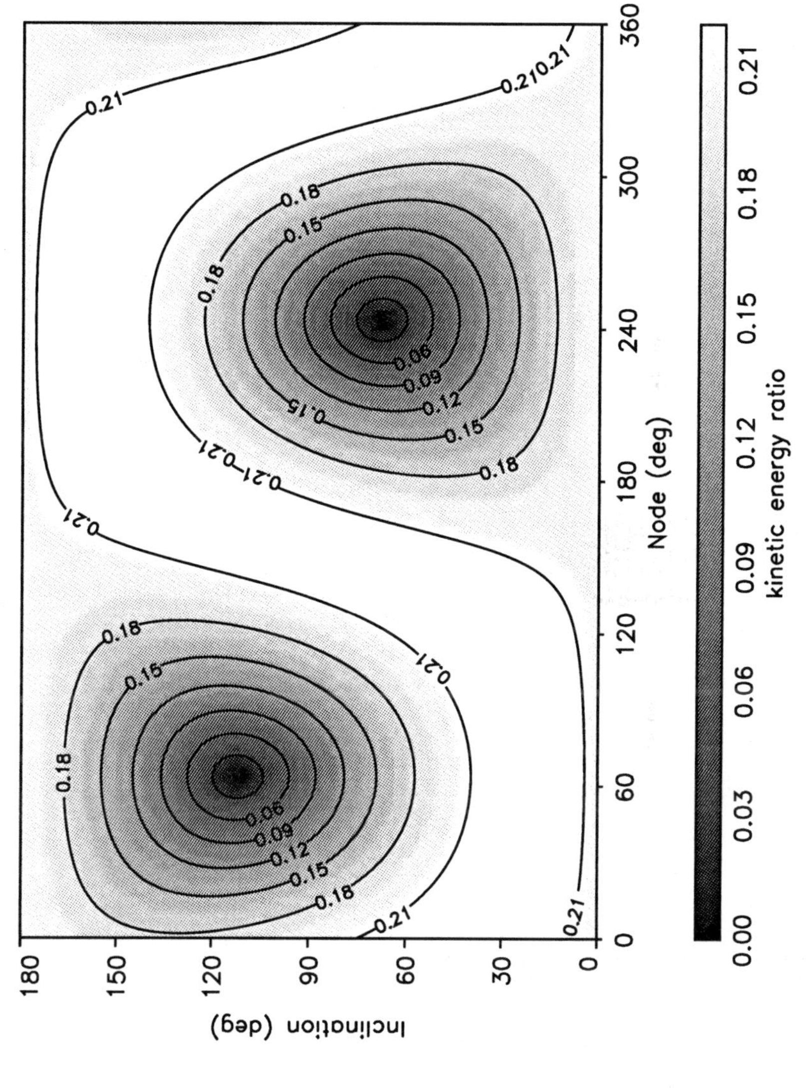


**Fig. 4.15. Kinetic energy ratio for MEO altitude orbits (altitude = 20,232 km).**

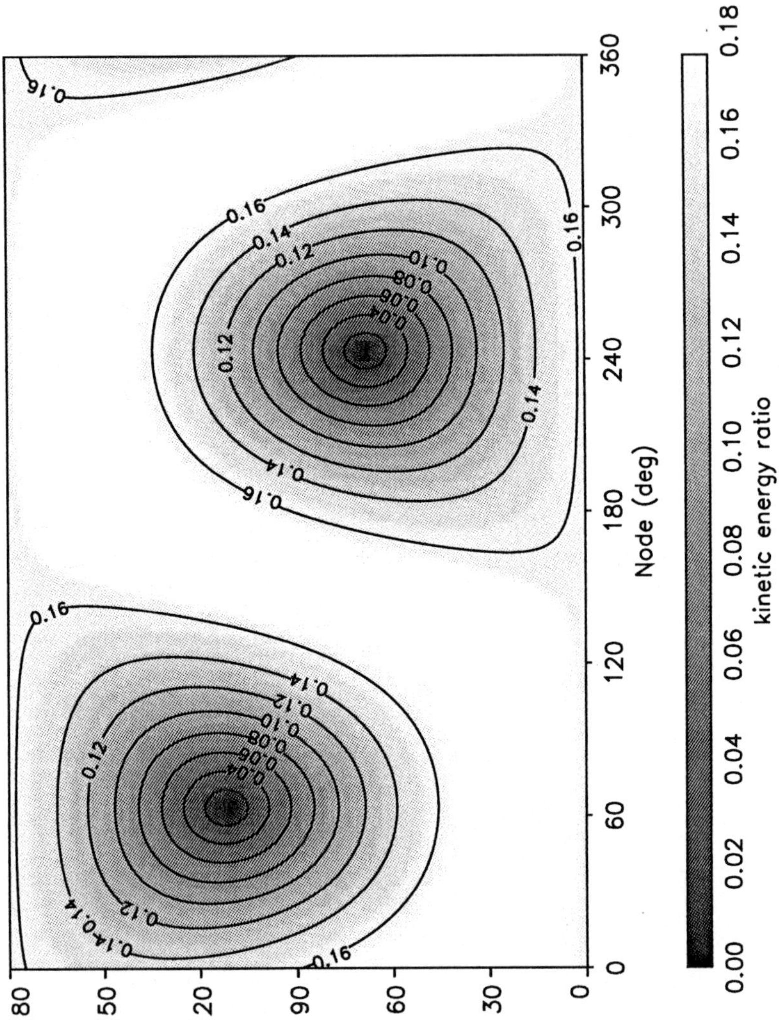


Fig. 4.16. Kinetic energy ratio for GEO altitude orbits (altitude = 35,862 km).

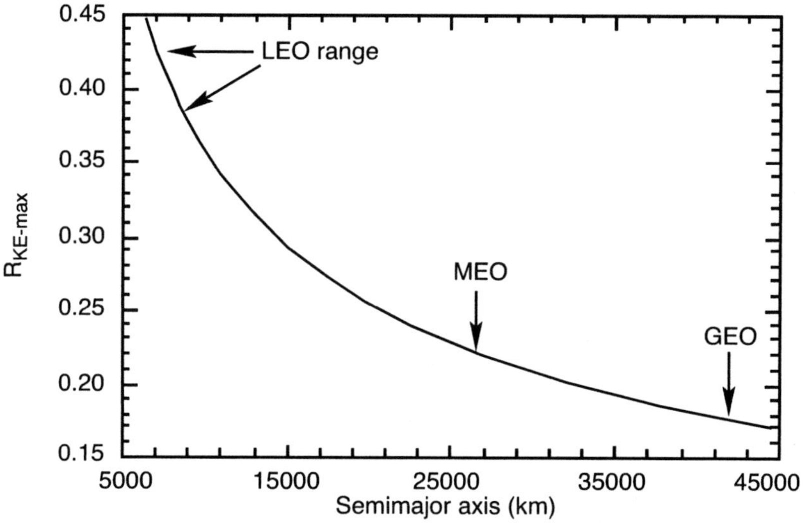


**Fig. 4.17. Maximum kinetic energy ratio as a function of semimajor axis.**

### 4.5.2 Molniya Orbits

For a Molniya orbit (HEO), the satellite under consideration is in a high-altitude orbit with a 12-hour period ($a$ ~26 610 km), set eccentricity ($e$ ~0.71), inclination ($i$ ~63.4 deg), and argument of perigee ($\omega$ ~270 deg). The spacecraft velocity of Eq. (4.52) for eccentric orbits in Cartesian geocentric coordinates is:

$$\bar{v}_s = v_m \begin{bmatrix} -\sin f \sin\Omega \cos i + (e + \cos f)\cos\Omega \\ \sin f \cos\Omega \cos i + (e + \cos f)\sin\Omega \\ \sin f \sin i \end{bmatrix} \tag{4.63}$$

where $f$ is the true anomaly (equal to the argument of latitude minus the argument of perigee, $f = u - \omega$) and $v_m$ is now the mean velocity given by:

$$v_m = \sqrt{\frac{\mu}{a(1-e^2)}} \tag{4.64}$$

The quantity to be minimized is still the relative velocity magnitude of Eq. (4.55), but now the speed of the orbit is no longer constant due to the high eccentricity. Therefore the expression to be minimized is:

$$\frac{\partial[(\bar{v}_s \cdot \bar{v}_s) - 2(\bar{v}_s \cdot \bar{V}_\infty)]}{\partial f} = 0 \tag{4.65}$$

The two dot products are Eqs. (4.66) and (4.67).

$$\bar{v}_s \cdot \bar{v}_s = v_m^2[1 + e^2 + 2e\cos f] \tag{4.66}$$

$$\begin{aligned} \bar{v}_s \cdot \bar{V}_\infty = v_m V_\infty[&-\sin f \cos i \cos\delta \sin(\Omega - \alpha) \\ &+ \sin f \sin i \sin\delta + \cos\delta\cos(\Omega - \alpha)(e + \cos f)] \end{aligned} \tag{4.67}$$

Substituting the dot products into Eq. (4.65), taking the derivative, and solving for the optimal true anomaly where the relative velocity magnitude would be either a maximum or a minimum gives:

$$\tan f = \frac{V_\infty[\sin\delta\sin i - \cos\delta\cos i\sin(\Omega-\alpha)]}{V_\infty\cos\delta\cos(\Omega-\alpha) - v_m e} \tag{4.68}$$

which reduces to Eq. (4.59) if the eccentricity goes to zero and if it is noted that the zero references for the argument of latitude for the near-circular case and the true anomaly in the Molniya case are 270 deg apart ($\omega = 270$ deg). Note that as in the near circular case, there will be two solutions located 180 deg apart—one a minimum and the other a maximum. But unlike the near circular case, the minima and maxima of the elliptical solution will not be symmetric in magnitude. The optimal node solution is the same here as it was for the near-circular case (Eq. [4.58]).

The relative velocity magnitude can be written as:

$$v_{\mathrm{rel}} = V_\infty\left(1 + \frac{v_s^2}{V_\infty^2} - 2\frac{\bar{v}_s \cdot \bar{V}_\infty}{V_\infty^2}\right)^{1/2} \tag{4.69}$$

In a manner similar to the near-circular case, the kinetic energy at impact can be expanded, but instead of expanding about the sum of the squares of the particle and satellite speeds $v_s^2 + V_\infty^2$, the expansion is carried out only about the square of the particle speed (because the satellite speed is not constant for elliptical orbits). The ratio between the minimum and maximum number of dangerous particles after expansion for elliptical orbits becomes:

$$R_{\mathrm{RV}} = 1 - \frac{N_{d-\min}}{N_{d-\max}} = 4\left\{\left[-\frac{3}{2}(1-s)\gamma_p\right]\left(\frac{\bar{v}_s \cdot \bar{V}_\infty}{V_\infty^2}\right)\right\} \tag{4.70}$$

The maximum and minimum kinetic energies possible can be approximated by assuming the expression in the square brackets is unity and that the orbit is edge-on with respect to the meteors, with the perigee and apogee velocities lined up with the particle velocity. The kinetic energy ratio becomes:

$$R_{KE_{\max}} = 4\frac{v_m}{V_\infty} - 2\frac{v_m^2}{V_m^2}(1+e^2) \approx 4\frac{v_m}{V_\infty} \tag{4.71}$$

which is more general than Eq. (4.62). That is, the maximum reduction possible is approximately four times the ratio of the mean velocity (circular or elliptic) to the particle speed. This provides a good approximate measure of the maximum benefit to be gained through use of this strategy (~31% for Molniya). The actual values of Eq. (4.70) will, however, depend upon the specific orbit being examined as well as the particular meteor stream (mass index $s$) and the type of shielding and damaging phenomena (velocity index $\gamma_p$).

Figure 4.18 shows the kinetic energy ratio for an HEO orbit without the fixed inclination of the Molniya. It is represented here for comparison with the preceding MEO plot (Fig. 4.15). Both of the plots are for the same mean altitude,

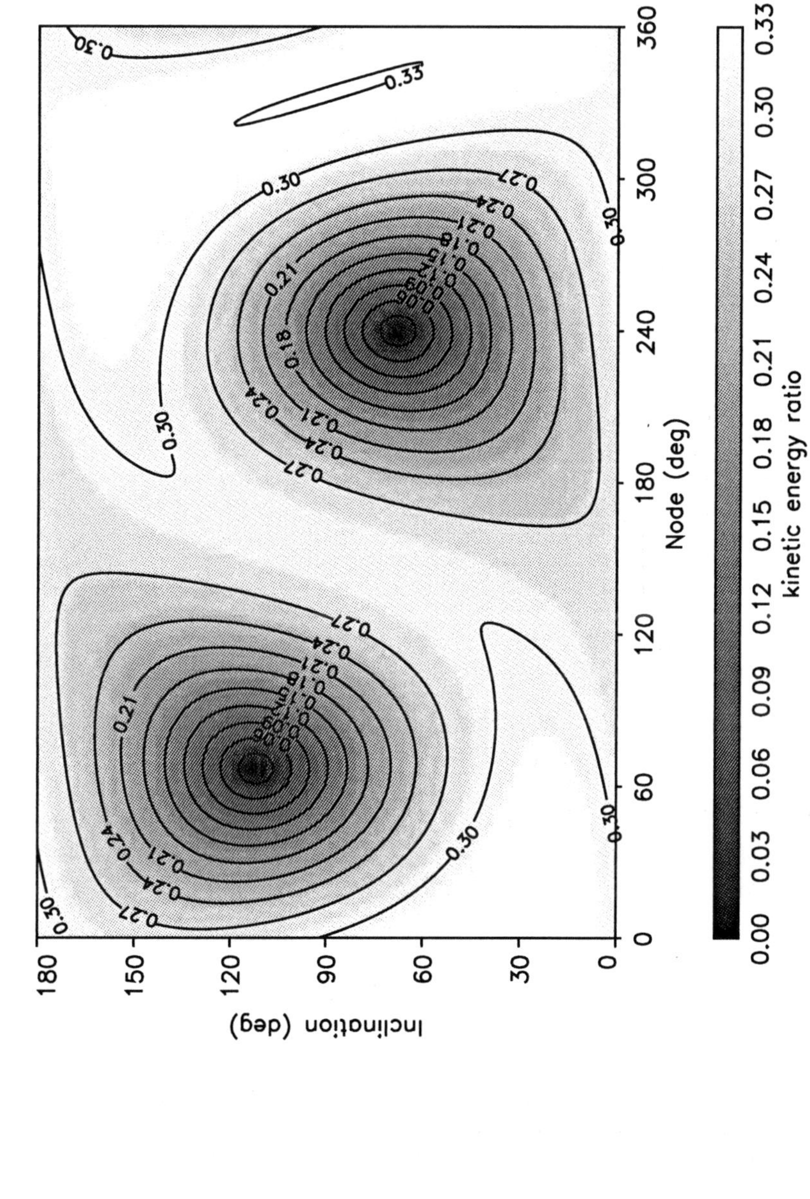


**Fig. 4.18. Kinetic energy ratio for HEO altitude orbits (mean altitude = 20,232 km).**

but different eccentricities. Overall, the higher the eccentricity, the steeper the gradient into the low regions where the ratio is close to zero. Additionally, the higher the eccentricity, the higher the overall peak.

Both the true anomaly for the minimum impact energy solution and the corresponding kinetic energy ratio are plotted in Fig. 4.19 for the specific Molniya orbit ($i$ ~63.4 deg) over the possible ranges of the right ascension of ascending node. The kinetic energy ratios range over a wide disparity from just over zero to ~33%, implying that a significant reduction in the impact energy is possible with the appropriate choice of the node and true anomaly.

The best node locations occur when the node is closely aligned with the right ascension of the meteor shower (153.45 deg and 333.45 deg). The match is not exact because of the fixed inclination and eccentricity; the actual optimums are ~147 deg and 344.5 deg. These optimums do not match the optimal node of Eq. (4.58) either (141 deg and 321 deg), implying that the best configuration for elliptical orbits is slightly offset from an edge-on orientation if the perigee is fixed. This offset is driven by the $v_s^2$ terms in Eq. (4.55), which disappear for near-circular orbits. The best along-track location for a satellite to be in for a Molniya at the time of the storm for the actual optimums is close to the perigee and apogee, respectively, at 23.7 deg and 156.9 deg. The worst node is close to the situation where the right ascension of the shower is nearly perpendicular to the node (~240 deg) with a corresponding true anomaly of ~270 deg.

### 4.5.3 Best and Worst Geographic Locations

The best and worst locations have been determined for individual satellites in terms of the orbital elements. But since the majority of satellites in HEO, MEO, and GEO orbits are tied to the geography of the Earth, it is useful to know where the best and worst geographical locations in the sense of the relative velocity

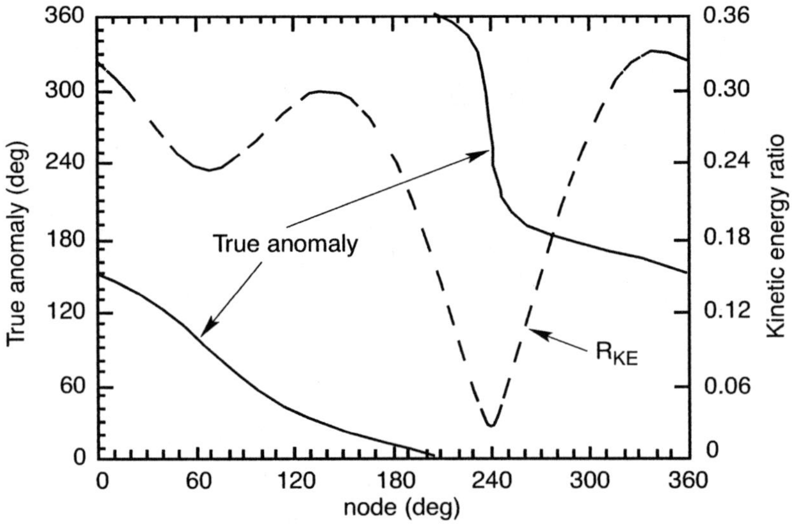


**Fig. 4.19. Molniya minimal in-plane location and kinetic energy ratio.**

magnitude will be during the Leonid storms of the next couple of years. Actual forecasts can be found from the orbital geometry for the specific times and best viewing locations (subradiant point) of the Leonid storms during this time (Table 4.2).The best viewing locations will be over the western Pacific and Asia in 1998, western Russia in 1999, and over the mid-Atlantic in 2000. The best satellite location (assuming an orbit that is locked to the longitude, i.e., a low-inclination prograde GEO) will be approximately above and 90 deg to the east of the best viewing location (this is a valid assumption within the level of error of the appearance of the storms), and of course the worst will be directly opposite of the best (Fig. 4.20: the Leonids approach the Earth almost perpendicular to the Earth–Sun direction; the best location moves away from the meteors, the worst into the approaching particles).

In 1998, satellites over the western United States and the eastern Pacific Ocean were in the best location (confirmed through observations), and satellites over eastern Europe and western Russia were in the worst location. In 1999, satellites over the western Pacific and Asia are in the best position traveling with the particles, while satellites over the United States are starting to travel into the meteor stream and so will suffer a greater impact energy. The best location for the year 2000 event will be over eastern Europe and Russia, with the worst location in the region over the eastern Pacific and the United States.

**Table 4.2. Leonid Visibility with Best and Worst Satellite Longitudes**

| Date (± 2 hours) | Best visibility (deg E. longitude) | Best (deg E. longitude) | Worst (deg E. longitude) |
|---|---|---|---|
| 1998 Nov 17: 20 UT | 130 ±30 | 220 | 40 |
| 1999 Nov 18: 03 UT | 36 ±30 | 126 | 306 |
| 2000 Nov 17: 10 UT | 313 ±30 | 63 | 243 |

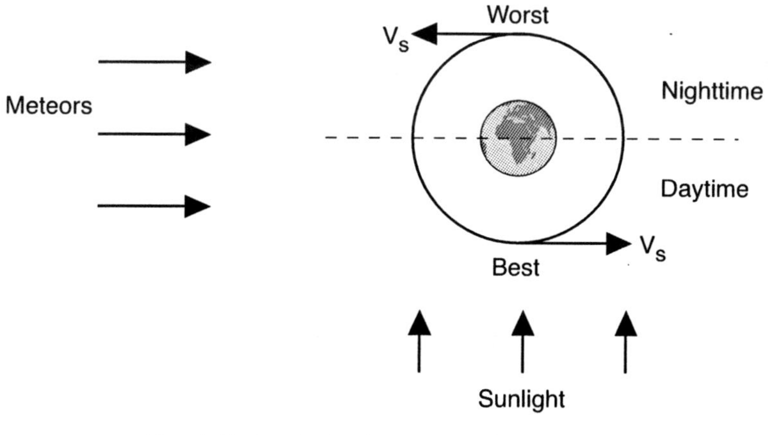


**Fig. 4.20. Leonid in-plane encounter geometry.**

### 4.5.4 DeltaV Analysis for Drifting Burns

Having determined from orbit propagation techniques where the optimal (i.e., minimal relative velocity magnitude) location is and knowing where the satellite will be at the time of the storm, the next issue of interest would be to determine just how expensive is it in terms of deltaV to move the satellite into the optimal in-plane location. Since LEO orbits are impractical and GPS must maintain its constellational integrity, the deltaV analysis that follows will be performed only for GEO orbits as an example. Other types of orbits can be examined as needed. Two types of maneuvers will be examined—one where time is not as important a consideration as minimizing the fuel, i.e., utilizing small drifting burns to place the satellite into the proper location, and the other where operational constraints force the shortest time off-line as possible, i.e., a larger impulsive burn to the lowest perigee (i.e., the lowest velocity relative to the meteors), followed by recovery to the original orbit. Only in-plane maneuvers will be considered here; out-of-plane maneuvers to alter the node are examined in Sec. 4.6.

The first maneuver is a simple four-part phasing maneuver (Fig. 4.21) performed early enough that the satellite has time to drift into the desired optimal location. If an initial phase difference between optimal location and the location of the satellite at the time of the storm is denoted as $\Delta u$, then the first part of the maneuver consists of correcting for $\Delta u$; the second consists of stopping the drift

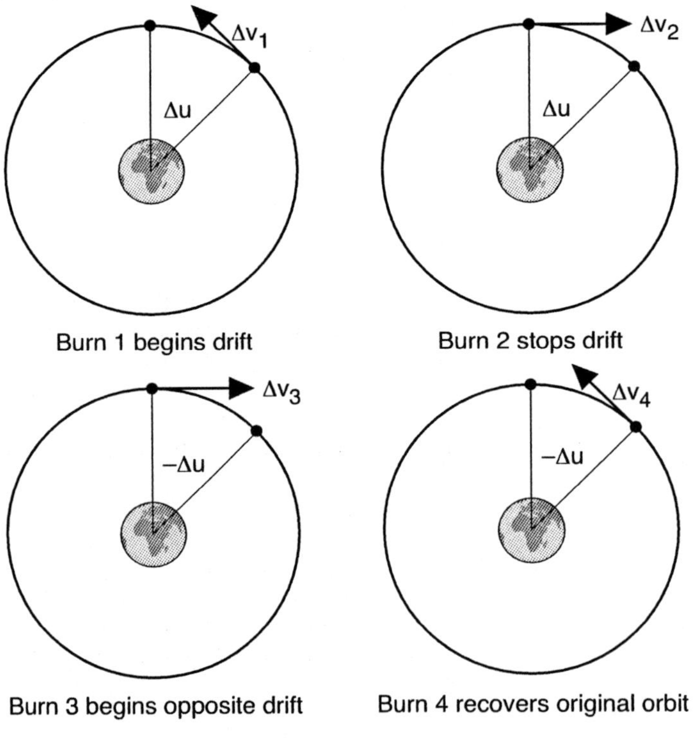


**Fig. 4.21. Four-part burn scenario for phase correction.**

in that direction. The storm is then allowed to pass. At this point, the satellite is $\Delta u$ behind where it should be, and a two-burn maneuver symmetrical to the first (but correcting in the other direction) must be performed to restore the original angular location in the orbit.

Assuming that the deltaV is small, the change in the mean motion (the drift rate) is (in deg/orbit, 1 orbit = 1 day for GEO [Wertz and Larson, 1991]) (subscript of 0 indicates original orbit):

$$DR = 360\frac{(n - n_0)}{n_0} \tag{4.72}$$

where the mean motion is given by:

$$n = \sqrt{\frac{\mu}{a^3}} \tag{4.73}$$

Expanding the mean motion in terms of a small change in the semimajor axis and substituting into the drift rate equation yields:

$$DR = 540\frac{\Delta a}{a_0} \tag{4.74}$$

The relative between a given deltaV and the subsequent change in the semimajor axis is (Eq. [4.94]):

$$\frac{\Delta a}{a_0} = 2\frac{\Delta v}{v_s} \tag{4.75}$$

where $v_s$ is the spacecraft velocity. This gives a final drift rate as a function of the magnitude of the burn as:

$$DR = 1080\frac{\Delta v}{v_s} \tag{4.76}$$

The drift rate can also be written in terms of the desired angle change and the $\Delta t$ over which the drift occurs (i.e., time before the storm when the burn is applied). Note that an important consideration is that, assuming the satellite's mission is geographically dependent, the longer $\Delta t$ is, the more mission downtime occurs (mission downtime will be twice the $\Delta t$). Therefore it is desirable to have not only the minimal deltaV, but a minimal $\Delta t$ as well. The relation is:

$$\Delta u = DR * \Delta t \tag{4.77}$$

Solving for the deltaV gives (where a factor of 4 has been included to account for the symmetries of the initial drift burn, recovery to the original orbit at the desired location, burn to set the drift into the opposite direction, and recovery to the original position in the original orbit) ($\Delta u$ is in degrees, $\Delta t$ is in days, $v_s$ is in km/sec):

$$\Delta v_{\text{total}} = 4\frac{\Delta u}{\Delta t}\frac{v_s}{1080} \tag{4.78}$$

Note that the expansion given here breaks down for large deltaVs. However, because errors of less than 10% accompany deltaVs up to 400 m/sec and the fuel consumed is desired to be a minimum, it is unlikely that large-enough deltaVs will be performed that would invalidate Eq. (4.78). Therefore, no loss of generality is caused by making the previous approximations.

Figure 4.22 shows the deltaV, time-before-storm at which the burn is applied, and the phase angle change desired. Units are in base 10 log of the deltaV for purposes of clarity. For fairly small deltaVs, significant changes in the argument of latitude can be generated if the burn is conducted far enough in advance of the storm. Compare these results with the deltaV requirements for changing the node shown in Fig. 4.37. This latter strategy takes advantage of Earth shadowing to protect the satellite and is discussed fully in Sec. 4.6. The difference in orders of magnitude between the fuel requirements of the two methods is apparent. However, the large times involved make the drifting burn strategy impractical for all but small changes in argument of latitude. If spares are being protected, then large mission downtimes are not a concern.

### 4.5.5 DeltaV Analysis for Impulsive Burns

In contrast to the small magnitude and large times necessary for drifting burns to be performed, a larger impulsive burn can be performed that will shorten the mission downtime. For example, if a GEO satellite is determined to be in the worst location possible in terms of the relative velocity magnitude at the time of the storm (headed into the storm), then a single burn can be applied one or more revs before the storm hits. Doing so will alter the shape of the orbit, with the result that the satellite will be in the optimal in-plane location at the time of the storm, moving in the same direction as will be the meteoroid particles. In the best situation possible, this point will also correspond to the perigee of the new orbit, with the result that the satellite's velocity will be at its largest as well as in the same direction as the meteors' velocity. Once phasing occurs where the satellite in the lower orbit resynchronizes with its original location, a second burn of equal magnitude is then applied at apogee of the new orbit in order to recover the original altitude. Figure 4.23 shows the mission profile for this strategy.

While the phasing requirements most likely will require multiple revs in the lower orbit, an approximate idea as to the deltaV requirements necessary to generate a particular level of decreased relative velocity magnitude can be found by assuming a near-circular orbit that is edge-on with respect to the approaching particles. Unlike the previous drifting burns, which minimized fuel consumption, the objective here is to have as little downtime as possible.

For example, if a GEO satellite will be in the worst location at the time of the storm, 24 hours in advance of the storm (i.e., 1 rev), the satellite can burn at just over 0.5 km/sec to place itself into a 16-hour period orbit. Such an action will place the satellite in a more favorable location at the time of the storm 24 hours later—going in the same direction as the particles, with a velocity of 4.8 km/sec. After another day passes, the satellite will again be in the original location at the proper time, and another 0.5 km/sec burn will take place to recover the original orbit. The total burn is just over 1 km/sec, with a mission downtime of 2 days. No other specific burn will accomplish the rephasing as quickly. There is a 20% reduction in the kinetic energy of impact between the GEO satellite heading into

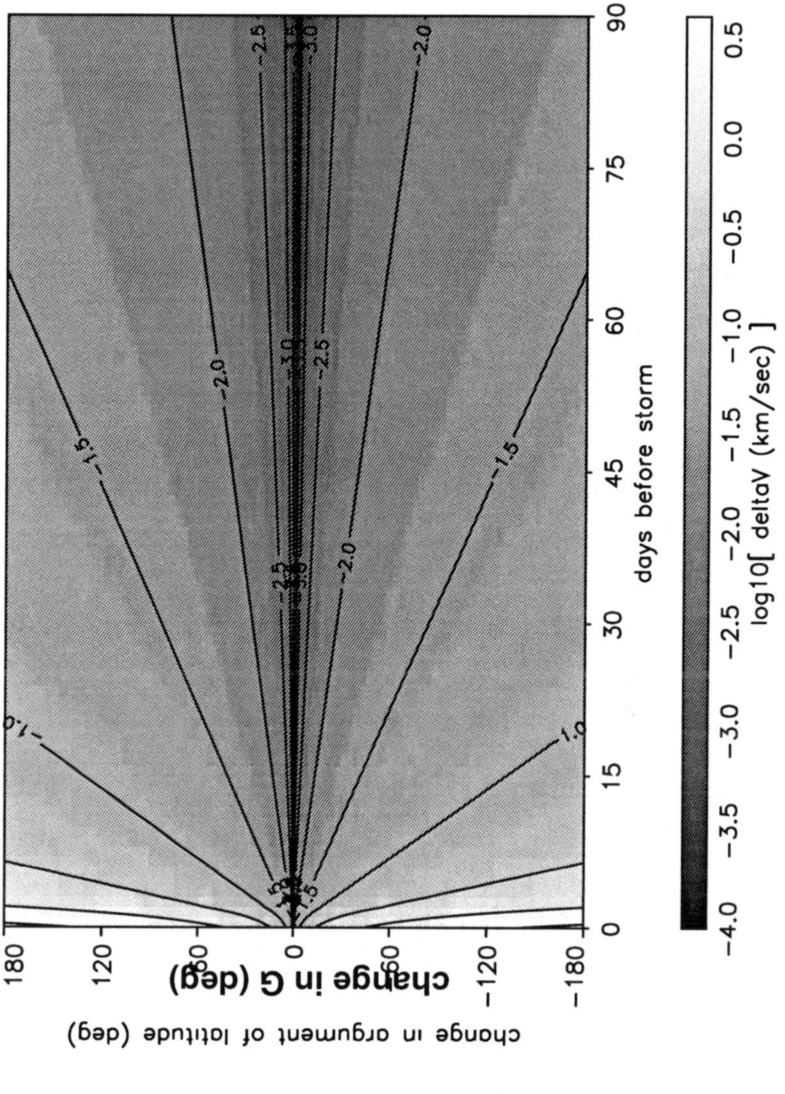


**Fig. 4.22. GEO deltaV requirements for given phase and burn time.**

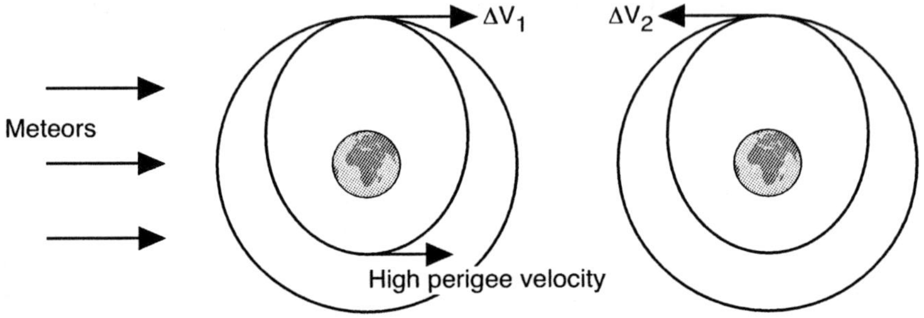


**Fig. 4.23. Two-part burn scenario for increased perigee velocity.**

the storm and the adjusted satellite headed away from the storm at perigee of the intermediate orbit. Lesser deltaVs to accomplish this are possible but require longer mission downtimes. Note that specific temporary orbits must be chosen in order to experience the recovery phase correctly. The drifting burns of Sec. 4.5.4 and the large impulsive burns described here represent the two extremes of possible in-plane maneuvering. Operational burns are most likely to be closer to the drifting burns than the large impulsive burns.

### 4.5.6 Constellations

This completes the analysis for individual satellites, but what about constellations? The benefit gained or lost through in-plane manipulation is minimal if the integrity of the constellation is to be maintained (i.e., every satellite in the plane is shifted). Consider for the sake of simplicity a single plane of a constellation that is edge-on to the approaching meteors (Fig. 4.24).

Regardless of the number of satellites in the plane, the total relative velocity of the constellation is simply the sum of the individual satellite relative velocities (where nsat is the number of the satellites in the constellation):

$$v_{rel-total} = \sum_{i=1}^{nsat} v_{rel-i} \tag{4.79}$$

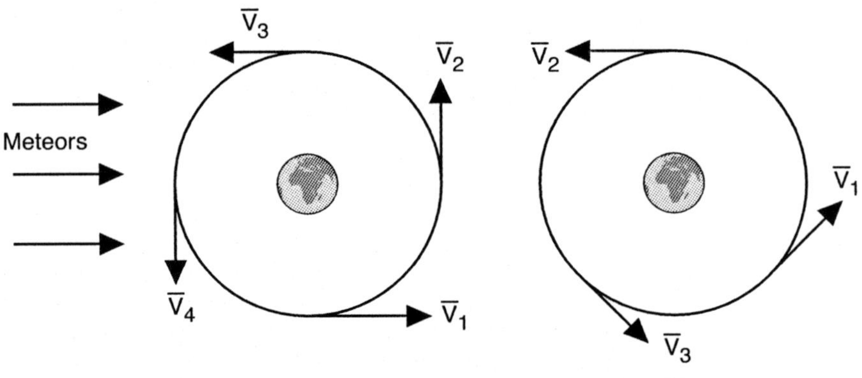


**Fig. 4.24. Sample geometry of constellation of satellites.**

If a near-circular orbit is assumed where the satellites all have the same velocity magnitudes (but different directions), the total impact energy becomes Eq. (4.80).

$$KE_{\text{total}} = \frac{1}{2}m_p\left[\text{nsat}(V_\infty^2 + v_s^2) - 2\bar{V}_\infty \cdot \sum_{i=1}^{\text{nsat}} \bar{v}_{s_i}\right] \tag{4.80}$$

If any optimization is to take place, it must therefore occur in the sum of the velocity vectors of the individual participants. But since the constellation is here assumed to be symmetric and the satellites evenly spaced in the orbit, then regardless of the number of satellites in the plane, the sum of the velocity vectors will be zero. This implies that no benefit is gained in terms of the relative velocity magnitude if the entire constellation is shifted, as long as the spacing between the satellites remains constant. However, the individual satellites in the constellation could still be moved in the plane of the orbit in order to gain the possible benefit for that particular member of the group, but at the cost of potentially disrupting the constellation's coverage. For HEO constellations, the sum of the velocity vectors would not necessarily be zero, but for MEO (GPS) and GEO, no relative velocity magnitude benefit is gained by shifting the entire constellation by a constant amount.

### 4.5.7 Limitations of the Analysis

A few final notes are in order about other types of meteor showers and the approximations used here. The analysis presented was for the Leonids, one of the fastest-moving meteor streams (~70 km/sec). However, other streams have produced large storms in the past, most notably the Draconids in the 1930s and 1940s with peak ZHRs over 10,000. But the Draconids (and other showers) are much slower than the Leonids, with a consequently lower impact energy. The relative velocity magnitude equations should therefore be used with caution because their expansion depends upon the meteor particle speed being much greater than the spacecraft speed. As these values become close, the first-order expansion will break down. Even so, the geometrical analysis holds for all showers regardless of their speeds, and the in-plane strategy can still be used to lower the impact energy for any meteor stream.

For example, the maximum kinetic-energy ratio of Eq. (4.71) gives an approximate overall gauge as to the usefulness of the in-plane methodology for any meteor shower with velocity $V_\infty$. For the slower-moving meteor showers (20 to 30 km/sec; the Draconids have a motion of ~23 km/sec, Geminids ~36 km/sec), this implies that the impact energy could almost be halved for a satellite at any altitude in the right combination of the node and inclination. Note that the approximation breaks down as the ratio approaches unity; a value of unity implies that the minimum kinetic energy is zero, or that the meteors are moving with the same speed in the same direction as the spacecraft. No meteor showers are known that move with speeds on the same order as spacecraft velocities (<~7 km/sec), therefore this is not a realistic scenario. A value greater than unity implies negative kinetic energy, which is not physically possible in any event. From a numerical analysis, it was found that the error grows significantly for values over 50%; therefore this regime should be treated with caution. For these events, a complete analysis of the impact energy without the expansion should be conducted.

It should be noted that the ballistic and plasma equations of Sec. 4.2.1 that drive the level of damage sustained through impact consist of the velocity magnitude raised to the power $\gamma_p$, while the kinetic energy goes as the magnitude squared. Therefore, while Eqs. (4.61) and (4.70) gave the benefit gained through use of this strategy in terms of the kinetic energy, this must be understood as a generalized result and specific meteor showers and spacecraft would need to be examined individually to determine the actual benefits.

Also, the benefit ratio equations presented elsewhere in this chapter all use the number of impacts that a spacecraft would experience if that vehicle were sitting stationary at the location of the Earth's center. In order to be consistent with the results from the other sections, the benefit ratios of Eqs. (4.61) and (4.70) should be computed using this same stationary reference point. The relative benefit gained from minimizing the relative velocity using this reference becomes logically enough one-half of the relative benefit ratio between the minimum and maximum points:

$$R_{RV} = \frac{N_0 - N_{\text{reduced}}}{N_0} = 2\left[-\frac{3}{2}(1-s)\gamma_p\right]\left(\frac{\bar{v}_s \cdot \bar{V}_\infty}{V_\infty^2}\right) \tag{4.81}$$

This can actually increase the benefit gained from the relative velocity minimization strategy over that given by Eq. (4.71), but is dependent upon the shield type and the phenomenon (i.e., spall or ESD). A summary of the relative velocity minimization methodology is given in Algorithm 4.4.

1. Optimal node location: $\sin(\Omega_{\text{opt}} - \alpha) = -\dfrac{\tan\delta}{\tan i}$

2. Optimal in-plane location: $\tan u_{\text{opt}} = \dfrac{\cos\delta\cos(\Omega-\alpha)}{\cos\delta\sin(\Omega-\alpha)\cos i - \sin\delta\sin i}$

3. DeltaV requirements and mission downtime are related to the amount of change needed to the satellite's in-plane location ($\Delta u$ is in degrees, $\Delta t$ and $TP$ are in days, velocities in km/s; this expression assumes a circular orbit with period $TP$):

$$\Delta v_{\text{total}} = \frac{8}{3}\frac{TP}{\Delta t_{\text{downtime}}}\frac{\Delta u}{360}v_s$$

$$\Delta v_{\text{total}} = 4\frac{\Delta u}{\Delta t}\frac{v_s}{1080}$$

4. Benefit gained expressed as a reduction in the damaging flux:

$$R_{RV} = \frac{N_0 - N_{\text{reduced}}}{N_0} = -3(1-s)\gamma_p\left(\frac{\bar{v}_s \cdot \bar{V}_\infty}{V_\infty^2}\right)$$

5. Applicability: GEO, less so for MEO and HEO. LEO cannot benefit from this strategy. Mission operation requirements may limit this strategy to on-orbit spares.

**Algorithm 4.4. Orbit maneuvering for relative velocity minimization.**

### 4.5.8 Combining In-Plane Strategies

The in-plane location strategies presented here for relative velocity magnitude minimization and previously for flux control (Sec. 4.4) can complement each other. Since the flux control is dependent upon the Earth-core vector and the velocity minimization is a function of the meteor radiant, the two options are independent of each other (Fig. 4.25).

Careful choice of the argument of latitude may therefore lead to a greater benefit than is possible from either strategy just by itself. Conversely, there may be arguments of latitude where one strategy yields good results while the other yields bad. Because the velocity minimization reduces the amount of damage caused from impact, while the flux control strategy reduces the probability of impact, the two relative benefits are difficult to compare and will vary from satellite to satellite and from event to event, depending upon mission constraints and operator preference. However, the reduction in the relative flux ratio that creates spall or ESD (Eq. [4.81]) can be compared with the reduction in the relative flux that straightforwardly impacts a satellite through the flux minimization strategy. This is shown in Fig. 4.26 for a near-equatorial GEO undergoing a Leonid event where the stream core is located on the sunward side. For the Leonids, then, the two in-plane strategies will partially counteract each other, depending upon the phenomenon being reduced (plasma, penetration, or impact probability) and just what the spatial distribution of the flux is. For other meteoric events (i.e., a Leonid streamlet core passing to the outside of the Earth's orbit), the two options may actually complement each other. Each case will need to be examined separately.

## 4.6 Time Minimization Using Earth Shadowing

Consider the equation describing the number of particles that impact on an exposed spacecraft (Eq. [4.24]):

$$N = \Phi A_{s/c} \Delta t_{\text{event}} \tag{4.82}$$

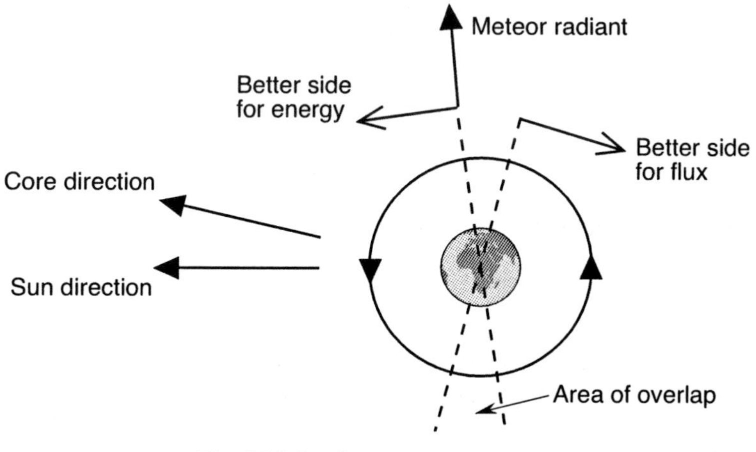


**Fig. 4.25. In-plane strategy overlap.**

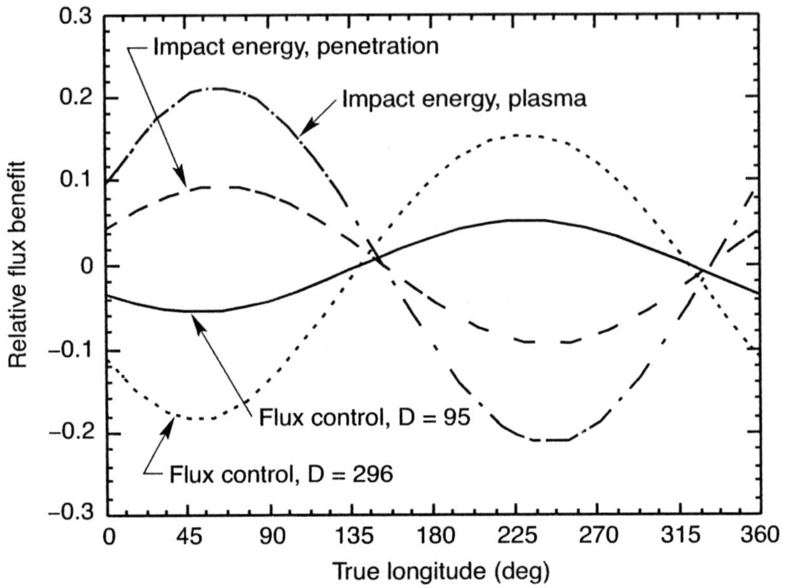


**Fig. 4.26. Leonid in-plane strategy overlap benefits for GEO.**

The number $N$ can be reduced through area minimization (Sec. 4.3) and by flux control (Sec. 4.4). The time of the event itself cannot be altered, but the amount of time the spacecraft spends exposed to the particles can be. This is done by taking advantage of the shadow cast by the Earth, the geometry of which was examined in Chapter 2. In this way, the spacecraft spends at least a portion of the storm sheltered. Chapter 2 examined the geometry of Earth shadowing, giving expressions for the optimal node location and the effectiveness (i.e., the amount of the orbit that is covered).

Since GEO satellites in near-equatorial orbits are not likely to gain any benefit from shadowing because of their high altitude (Sec. 2.1.2), taking advantage of Earth shadowing is most effective for the lower-altitude LEO- and possibly MEO-type satellites. However, large uncertainties exist in the timing of any potential storm; therefore, the in-plane location of LEO and MEO satellites will not be predictable to a large extent. As a consequence, it is not the in-plane location that determines the level of protection, but the orientation of the orbit. If a portion of the orbit plane is behind the Earth, the probability of impact is reduced even though the exact time of the storm is not well known.

The optimal situation is achieved when the orbit passes directly behind the Earth with respect to the incoming meteors. In other words, the orbit plane is oriented to contain the meteor's approach vector. However, it is unlikely that this situation will naturally occur, and the orbit will not be in the beneficial geometry when the meteor particles hit the vicinity of the Earth. It may therefore be desirable for mission operators to move the orbit into the proper orientation. This section examines the deltaV requirements necessary to move the orbit plane of a given satellite into the proper location in order to take advantage of the benefits of Earth shadowing. Three scenarios are examined:

- Moving the node through a direct burn (the fastest and most expensive means possible)
- Altering the orbit plane through a lunar flyby (cheaper but slower)
- Changing the orbit elements so that the natural secular precession in the node owing to the Earth's oblateness makes the desired change in the node (cheapest but takes the most time)

### 4.6.1 DeltaV Requirements for Direct Burns

The optimal node location will occur when the orbit plane contains the meteor approach vector ($\alpha$, $\delta$); in this case the dot product between the vector perpendicular to the orbit and the approach vector goes to zero, giving the desired nodal relation between the node and inclination (Sec. 2.1):

$$\sin(\Omega - \alpha) = -\frac{\tan\delta}{\tan i} \tag{4.83}$$

Although Eq. (4.83) gives the location of the optimal node, this section is concerned with orbit planes that are not naturally in this orientation and the subsequent deltaV requirements necessary to move them there. The first situation to consider is simply a direct movement of the right ascension of ascending node to the desired location through a single burn. To move the node through a desired angle $\Delta\Omega$, the deltaV requirements are computed for near-circular orbits as (Vallado, 1997):

$$\Delta v_{\Omega\mathrm{only}} = 2v_s \sin\left(\frac{\vartheta}{2}\right) \tag{4.84}$$

where $v_s$ is the spacecraft velocity and $\vartheta$ is the burn angle defined through spherical trigonometry as:

$$\cos\vartheta = \cos^2 i + \sin^2 i \cos\Delta\Omega \tag{4.85}$$

The deltaVs required to forcibly move the node angle a significant amount are therefore prohibitive in terms of mission operations. Figure 4.27 shows the ratio of the deltaV required to move the node, compared with the circular velocity of the orbit for different values of the inclination in 15-deg increments, from 15 to 90 deg. As the ratio approaches unity, the required deltaV matches the orbital velocity itself. For highly inclined orbits, the required deltaV can approach twice the orbital velocity. Thus, moving the node directly so that the majority of the orbit lies in the Earth's shadow with respect to the incoming meteors is not a practical option other than for very small node changes. In addition, it must be noted that Fig. 4.27 gives only the first part of the maneuver. If the node is changed to accommodate the storm, after the storm has passed, it must be moved back through an equally sized angle to the original location. The total deltaV requirements are thus twice those shown here.

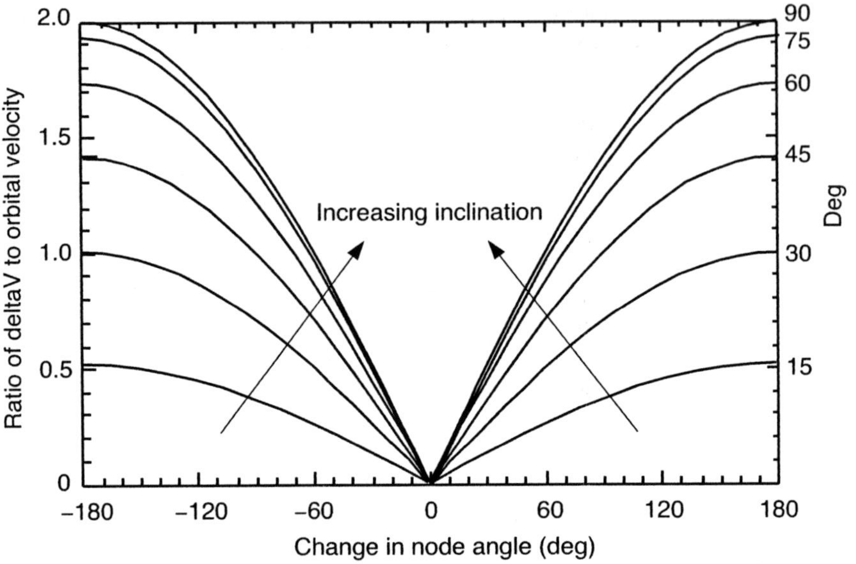


**Fig. 4.27. DeltaV requirements for direct node changes.**

### 4.6.2 DeltaV Requirements for Lunar Flybys

Another method can be utilized to change the orbit plane—boosting the satellite to the Moon and then using a hyperbolic flyby to change the Earth-centered plane into the desired configuration. The AsiaSat 3 spacecraft performed this type of maneuver successfully in 1998. The scenario involves burning to achieve an apogee equal to the Moon's altitude ($R_{em}$ ~384,400 km), flying by the Moon to change the plane, then reburning at perigee to circularize into the shadowed orbit. For simplicity, assume that the satellite is in a near-circular orbit with velocity $v_s$ and semimajor axis $a$, and that phasing requirements are ignored. Then the intermediate orbit can be characterized in terms of the energy equation (Eq. [4.92]) as a function of the deltaV necessary to achieve lunar altitude:

$$\Delta v_{\text{lunar}} = \frac{4}{v_s}\left(\frac{\mu}{2a} - \frac{\mu}{R_{\text{em}}}\right) \tag{4.86}$$

Figure 4.28 shows the normalized deltaV (ratio of deltaV to orbital velocity) as a function of the satellite's original semimajor axis. The fuel requirements for this type of maneuver are thus lower than the direct burns. However, it must be noted that this scenario also requires recovery burns of equal magnitude. In essence, the first burn is performed to boost to the Moon; a flyby of the Moon is performed to change the node into the desired location; then a second flyby is conducted to change the plane back; and a second burn is finally utilized to recircularize into the original orbit. In order to recover the original orbit, a factor of two thus must be considered in examining Fig. 4.28. Therefore, deltaVs on the order of the orbital speed are required to utilize this strategy. Using the

Moon to change the orbit plane is, like direct node angle changes, prohibitive in terms of the deltaV except in the most dire of circumstances.

Another possibility exists in using the Moon as a shield itself, namely, boosting the satellite out close to lunar altitude and letting the vehicle hide in its shadow. However, the satellite will be traveling (assuming the apogee is at lunar distance) at ~0.5 km/sec. Given a lunar radius of 1738 km, this implies that the satellite can spend at most about 2 hours in the shadow of the Moon. Since most storms are several hours in duration with associated levels of uncertainty, the protection afforded by a deliberate campaign of lunar shadowing is limited.

### 4.6.3 Dynamics of Earth Oblateness Burns

Another way of affecting the location of the node is to change the secular node rate in advance and let the natural nodal precession do most of the work. The Earth oblateness causes a secular perturbation in the node, forcing the node to continuously drift in a linear fashion. The node location at the time of the storm is therefore given in terms of the current node and the time to the storm by:

$$\Omega_{\text{storm}} = \Omega_{\text{current}} + \dot{\Omega}_{\text{current}}(t_{\text{current}} - t_{\text{storm}}) \tag{4.87}$$

where $\dot{\Omega}$ is the preburn value of the secular nodal drift caused by $J_2$. Therefore, by changing the secular node rate far enough in advance, the node location at the time of the storm can be made equal to the desired optimal location:

$$\Omega_{\text{optimal}} = \Omega_{\text{current}} + (\dot{\Omega}_{\text{preburn}}(t_{\text{current}} - t_{\text{burn}}) + \dot{\Omega}_{\text{postburn}}(t_{\text{burn}} - t_{\text{storm}})) \tag{4.88}$$

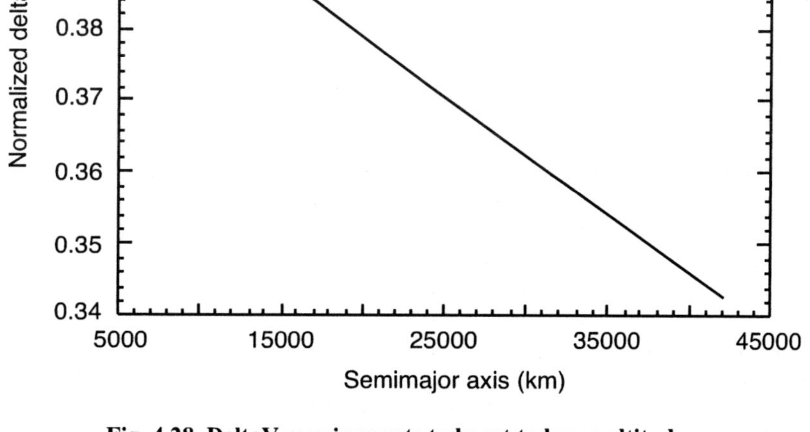


**Fig. 4.28. DeltaV requirements to boost to lunar altitude.**

To summarize, a satellite exists in a certain orbit of known node location with a secular rate dependent upon the orbital elements (Eq. [4.87]), but the resulting node location at the time of the storm will not be in the desired optimal location. Therefore, at some intermediate time a burn will occur that will change the orbital elements and hence the secular node rate such that the new node location at the time of the storm will now be in the optimal position (Eq. [4.88]).

The node motion is driven by the secular perturbation owing to the Earth's oblateness:

$$\dot{\Omega} = -\frac{3}{2}J_2 n\frac{\cos i}{(1-e^2)^2}\left(\frac{R_e}{a}\right)^2 \tag{4.89}$$

where $R_e$ is the Earth's radius and $n$ is the mean motion.

Although the secular drift in the node caused by the Earth's oblateness is most sensitive to changes in the inclination, these same inclination changes are most expensive in terms of deltaV. Therefore it is necessary to compare the various options available (i.e., changing $a$, $e$, or $i$) in terms of the required deltaV in order to find the most efficient way of altering the node rate.

Three orbit element changes affecting the secular node rate can result from a small impulsive burn—alterations in semimajor axis ($\Delta a$), inclination ($\Delta i$), and eccentricity ($\Delta e$). Binomially expanding to first order, the node rate in each of these orbit elements gives the adjusted node rate caused by an impulsive burn:

$$\dot{\Omega}_{\text{postburn}} = \dot{\Omega}_{\text{preburn}}\left(1-\frac{7}{2}\frac{\Delta a}{a}+\frac{4}{1-e^2}e\ \Delta e-\frac{\sin i}{\cos i}\Delta i\right) \tag{4.90}$$

By substituting this expression into Eq. (4.88) and taking the difference between Eqs. (4.87) and (4.88), the resulting change in the node rate becomes:

$$\Delta\dot{\Omega} = \frac{\Omega_{\text{optimal}}-\Omega_{\text{storm}}}{(t_{\text{burn}}-t_{\text{storm}})} = \dot{\Omega}_{\text{preburn}}\left(-\frac{7}{2}\frac{\Delta a}{a}+\frac{4}{1-e^2}e\ \Delta e-\frac{\sin i}{\cos i}\Delta i\right) \tag{4.91}$$

Therefore, for a given required nodal change and in order to keep the time of burn as close as possible to the time of storm (i.e., keeping the satellite's orbit in its operational status as long as possible), this combination of orbit element changes must be a maximum.

Each of the orbit element changes relevant to the secular node rate can be expressed in terms of a given deltaV. The semimajor axis change is derived from the energy of the orbit, the eccentricity from the angular momentum, and the inclination from simple geometry. The specific energy (energy per unit mass) of the orbit is simply the sum of the kinetic and potential specific energies:

$$\mathcal{E} = \frac{\bar{v}\cdot\bar{v}}{2}-\frac{\mu}{r} = -\frac{\mu}{2a} \tag{4.92}$$

The post-burn specific energy is:

$$\mathcal{E}' = \frac{(\bar{v}+\Delta\bar{v})\cdot(\bar{v}+\Delta\bar{v})}{2}-\frac{\mu}{r} = -\frac{\mu}{2(a+\Delta a)} \tag{4.93}$$

Because the impulsive burn is assumed to be small, higher-order terms can be neglected. Solving, then, for the change in semimajor axis gives:

$$\frac{\Delta a}{a} = \frac{2a}{\mu}(\bar{v} \cdot \Delta\bar{v}) \tag{4.94}$$

The most efficient change in semimajor axis will thus occur under two conditions—when the velocity is largest in the orbit (i.e., perigee), and when the impulsive burn is applied in the same direction as the orbital velocity vector.

The change in the eccentricity due to an impulsive burn can be found in a similar manner. As was the case for the semimajor axis, it is typical to adjust the eccentricity through an impulsive burn in an in-plane maneuver. It is most convenient to use the orbital angular momentum in computing the change in eccentricity:

$$\bar{h} = [\bar{r} \times \bar{v}] \tag{4.95}$$

The magnitude of the angular momentum can be written in terms of the orbit elements as:

$$(\bar{h} \cdot \bar{h}) = \|\bar{r} \times \bar{v}\|^2 = \mu a(1 - e^2) \tag{4.96}$$

Adding in the impulsive burn alters the angular momentum, which in turn causes changes in the semimajor axis as well as the eccentricity:

$$(\bar{h} + \Delta\bar{h}) \cdot (\bar{h} + \Delta\bar{h}) = \|\bar{r} \times (\bar{v} + \Delta\bar{v})\|^2 = \mu(a + \Delta a)[1 - (e + \Delta e)^2] \tag{4.97}$$

Expanding Eq. (4.97), substituting Eq. (4.94) for the semimajor axis change, neglecting higher-order terms, and rearranging gives the change in eccentricity from a small impulsive burn as:

$$e\,\Delta e = \frac{a(1 - e^2)}{\mu}(\bar{v} \cdot \Delta\bar{v}) - \frac{1}{\mu a}(\bar{h} \cdot \Delta\bar{h}) \tag{4.98}$$

The inclination change occurs most efficiently as a result of an out-of-plane burn at the node crossing (either ascending or descending) without changing the overall magnitude of the orbital velocity, just its direction. The three vectors (preburn velocity, postburn velocity, and impulsive maneuver) form an isosceles triangle, and it is a simple matter to compute the plane change as:

$$\frac{\Delta v_N}{v_T} = 2\sin\frac{\Delta i}{2} \tag{4.99}$$

where $v_T$ is the velocity component in the transverse direction and $\Delta v_N$ is the burn component normal to the orbit plane. Large plane changes thus require large amounts of deltaV. The most efficient inclination burn will therefore occur at the node that has the smaller transverse velocity. If it is again assumed that the burn is small in comparison to the velocity of the orbit, the resulting inclination change will be small as well and Eq. (4.99) can be approximated by:

$$\Delta i = \frac{\Delta v_N}{v_T} \tag{4.100}$$

If the impulsive burn is taken to be in an arbitrary direction at the node, an arbitrary burn can be written in terms of the radial, transverse, and normal coordinate system as:

$$\Delta\bar{v} = \Delta v\begin{bmatrix}\sin\phi\cos\theta\\ \cos\phi\cos\theta\\ \sin\theta\end{bmatrix} \tag{4.101}$$

where $\phi$ is the in-plane angle for the burn referenced to the transverse direction and $\theta$ is the out-of-plane (normal) component.

If this expression is substituted along with the conic equation of the orbit:

$$r = \frac{p}{1 + e\cos f} \tag{4.102}$$

and the spacecraft velocity written in the same RTN system:

$$\bar{v} = \sqrt{\frac{\mu}{p}}\begin{bmatrix}e\sin f\\ 1 + e\cos f\\ 0\end{bmatrix} \tag{4.103}$$

into the equations for the orbit element changes (Eqs. [4.94], [4.98], and [4.100]), those changes become:

$$\frac{\Delta a}{a} = 2\frac{\Delta v}{1 - e^2}\sqrt{\frac{p}{\mu}}\{e\sin f\sin\phi\cos\theta + (1 + e\cos f)\cos\phi\cos\theta\} \tag{4.104}$$

$$e\Delta e = \Delta v\sqrt{\frac{p}{\mu}}\left\{e\sin f\sin\phi\cos\theta + (1 + e\cos f)\cos\phi\cos\theta - \frac{1 - e^2}{1 + e\cos f}\cos\phi\cos\theta\right\} \tag{4.105}$$

$$\Delta i = \Delta v\sqrt{\frac{p}{\mu}}\frac{\sin\theta}{1 + e\cos f} \tag{4.106}$$

where $f$ is the true anomaly and $p$ is the semilatus rectum.

Substituting Eqs. (4.104), (4.105), and (4.106) into Eq. (4.91) gives the normalized node rate change for an arbitrary burn in three dimensions (normalized in this sense of the preburn node rate, Eq. [4.89]):

$$\Delta\dot{\Omega} = \Delta v\sqrt{\frac{p}{\mu}}\left\{\left(\frac{-3}{1 - e^2}\right)[e\sin f\sin\phi\cos\theta + (1 + e\cos f)\cos\phi\cos\theta] - \frac{4}{1 + e\cos f}\cos\phi\cos\theta - \frac{\tan i}{1 + e\cos f}\sin\theta\right\} \tag{4.107}$$

To find the burn angles and location in the orbit to apply the burn, it is a simple matter to take the derivatives of Eq. (4.107) with respect to the true anomaly $f$, in-plane burn angle $\phi$, and out-of-plane burn angle $\theta$, and set them equal to zero:

$$\frac{\partial \Delta\dot{\hat{\Omega}}}{\partial \phi} = 0 = -3(1 + e\cos f)e\sin f\cos\phi + [3(1 + e\cos f)^2 + 4(1 - e^2)]\sin\phi \tag{4.108}$$

$$\frac{\partial \Delta\dot{\hat{\Omega}}}{\partial f} = 0 = -3(1 + e\cos f)^2[\cos f\sin\phi - \sin f\cos\phi]\cos\theta - 4(1 - e^2)\sin f\cos\phi\cos\theta - (1 - e^2)\tan i\sin f\sin\theta \tag{4.109}$$

$$\frac{\partial \Delta\dot{\hat{\Omega}}}{\partial \theta} = 0 = 3(1 + e\cos f)[e\sin f\sin\phi + (1 + e\cos f)\cos\phi]\sin\theta + 4(1 - e^2)\cos\phi\sin\theta - (1 - e^2)\tan i\cos\theta \tag{4.110}$$

Consider for a moment the in-plane burn solution (i.e., the out-of-plane angle is constrained to be equal to zero). In this case, Eq. (4.110) has no meaning. By inspection of Eqs. (4.108) and (4.109), the solution to this case is to have the sines of the true anomaly and in-plane angle both equal to zero (this has been confirmed numerically as the only solution). This solution thus consists of the first half of a Hohmann transfer. For in-plane burns, a further numerical analysis confirms that the Hohmann transfer burn occurring at apogee is the best possible solution. Making the apogee substitutions into Eq. (4.91) gives:

$$\Delta\dot{\hat{\Omega}}_{\text{Hohmann - apogee}} = \sqrt{\frac{p}{\mu}}\Delta v\left(\frac{-7 - e}{1 - e^2}\right) \tag{4.111}$$

Similarly to the in-plane burn-only case, the most obvious solution when out-of-plane burns are considered is also to have the sines of the true anomaly and in-plane burn angle both equal to zero. Physically, this implies that the periapse is located at the node. Although it is a rather specialized situation, it does indicate that the Hohmann burn is optimal when considering in-plane burns only and does not necessarily provide the best solution possible when analyzing burns with out-of-plane components as well. In this case (perigee and apogee at the nodes) both Eqs. (4.108) and (4.109) are directly satisfied, and Eq. (4.110) can then be solved directly to obtain the optimum out-of-plane angle:

$$\tan\theta_{opt} = (\pm 1)_{\phi}\left(\frac{(1 \mp e)_f \tan i}{(7 \mp e)_f}\right) \tag{4.112}$$

The plus/minus signs correspond to the senses of the true anomaly (i.e., apogee or perigee) and the in-plane burn angle (i.e., forward or backward).

Note that Eq. (4.112) assumes the burn takes place at the node. For the near-circular orbits where the perigee is ill-defined, the orbit orientation is not an issue because the orbit is nearly circular and as a consequence the velocity (in RTN coordinates) is nearly the same throughout the orbit. Therefore, the difference between a true perigee/apogee burn reduces as the eccentricity reaches zero, and Eq. (4.112) would still apply. However, for elliptical orbits, it is not likely that the perigee or apogee (i.e., when the sine of the true anomaly is zero) will occur at the node crossings. In this case the burn can still take place at the node, but the true anomaly at the node crossing is not equivalent to the perigee or apogee. Equation (4.109), then, is no longer applicable (the true anomaly at node is allowed to be an arbitrary value), and Eqs. (4.108) and (4.110) can be used to solve for the optimal burn angles:

$$\tan\phi_{opt} = \frac{3(1+e\cos f)e\sin f}{3(1+e\cos f)^2+4(1-e^2)} \tag{4.113}$$

$$\tan\theta_{opt} = \frac{(1-e^2)\tan i}{3(1+e\cos f)[e\sin f\sin\phi+(1+e\cos f)\cos\phi]+4(1-e^2)\cos\phi} \tag{4.114}$$

Figures 4.29 and 4.30 give the optimum in-plane and out-of-plane burn angles for the elliptical case constrained to burn at the node (eccentricities range from 0.05 to 0.30 in increments of 0.05). While Fig. 4.29 shows the optimum in-plane angle, Fig. 4.30 gives a pseudo-out-of-plane angle. In order to reduce the need for multiple plots for various inclinations, the tangent of the inclination was assumed to be equal to 1 in Eq. (4.114). Thus the tangent of the out-of-plane angle shown in Fig. 4.30 needs to be scaled by the tangent of the inclination for

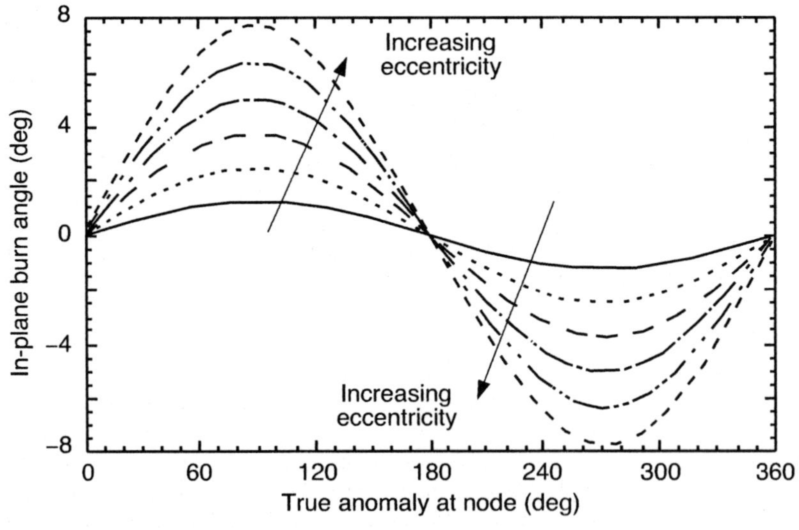


**Fig. 4.29. Optimal in-plane burn angle for arbitrary node burns.**

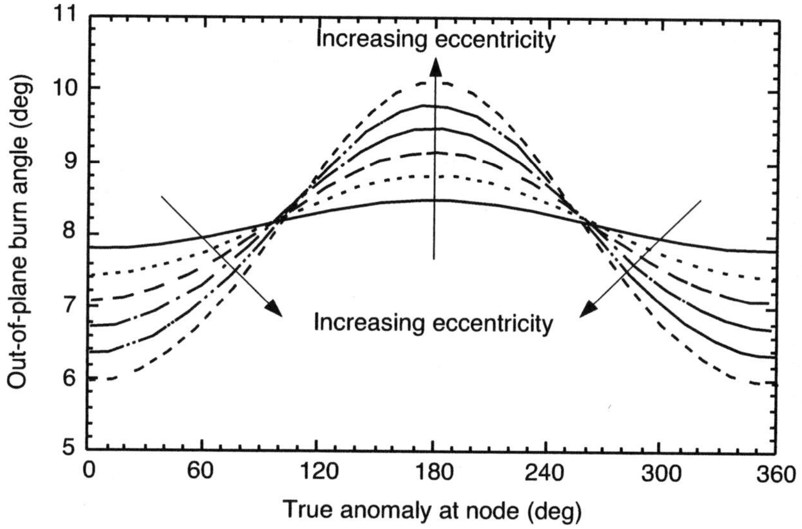


**Fig. 4.30. Optimal out-of-plane burn angle for arbitrary node burns.**

the specific orbit under consideration in order to be valid. However, the qualitative behavior of the out-of-plane burn angle can be seen from the diagram.

The nature of these solutions is that for any given orbit, the node crossing will occur at either of two locations—the ascending or descending nodes. These correspond respectively to the true anomaly being equal to the opposite of the argument of perigee and 180 deg opposite to that:

$$f = -\omega \text{ (ascending node)} \qquad f = 180 - \omega \quad \text{(descending node)} \qquad (4.115)$$

Therefore, there will be two general types of solutions, one corresponding to each of the nodes. The four node-specific solutions themselves (depending upon the in-plane and out-of-plane burn angles) are again symmetrical in nature. Figure 4.31 shows an example of the eight solutions for an eccentricity of 0.30 and inclination of 57 deg (doubly normalizing Eq. [4.107] to remove the semilatus rectum dependence), which gives a clear indication of the symmetry involved. With crossover points between the ascending and descending node curves at 90 deg and 270 deg, Fig. 4.31 indicates that whichever node the periapse is farther from is the better node to burn at. In other words, the node crossing that is further away from the Earth is the better choice for this type of maneuver. When the periapse point is closer to the ascending node, the burn should take place at the descending crossing; when the periapse is closer to the descending node, then the ascending node should be the maneuver point.

This solution describes the optimal case when the burn is constrained to be at the node in order to take advantage of any inclination alterations, but the location of the periapse is allowed to roam wild and free. Reexamine Figs. 4.29 and 4.30. Note that as the eccentricity reaches zero, the optimal in-plane burn angle reaches zero as expected (Fig. 4.29) and the out-of-plane burn angle approaches approximately 8 deg (recall in Fig. 4.30 that the tangent of the inclination was set

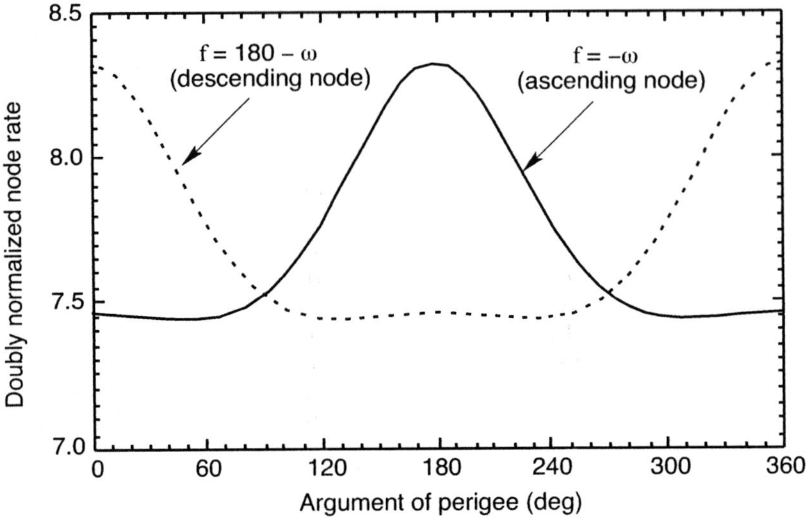


**Fig. 4.31. Symmetries in the full burn solution (e = 0.3, i = 57 deg).**

to 1 for this plot). This equates to the optimal out-of-plane solution when the line of apsides was constrained to be at the node crossing with a 45-deg inclination. But how does the optimal node-burn solution compare with the optimal in-plane solution (i.e., one-half of a Hohmann transfer) derived earlier? By comparing Eq. (4.111) for the apogee burn and Eq. (4.107) with the optimal node-burn angles (Eqs. [4.113] and [4.114]), this comparison can be examined.

Figure 4.32 shows the ratio of the optimal node burn to the apogee (in-plane optimal, Eq. [4.111]) solution for an eccentricity of 0.05 for the available range of inclinations. When the ratio is greater than unity, the optimal node burn becomes more effective than the apogee burn. It can be seen that as the true anomaly at node (i.e., at the argument of perigee) approaches 180 deg (i.e., when apogee is close to the node), the ratio becomes larger than 1, and at this point the node burn becomes more effective at changing the secular node rate due to Earth's oblateness than the Hohmann transfer burn at apogee. Also, at an inclination of 51 deg for the eccentricity shown here (0.05), the node burn is better for all the values of the argument of perigee.

As a consequence, for each eccentricity and true anomaly at node (i.e., argument of perigee), there will exist a potential critical inclination where the node burn becomes more effective than the apogee burn. The critical inclination for the node burn case with both in-plane and out-of-plane components can be solved for from Eqs. (4.111), (4.107), and (4.114) (where the optimal out-of-plane angle is also dependent upon the optimal in-plane angle, Eq. [4.113]):

$$\tan i_{cr} = \frac{(-7-e)}{(1-e^2)}(1+e\cos f)\sin\theta_{opt} \qquad (4.116)$$

This is shown in Fig. 4.33, where the regions denoting a more efficient burn at the node are above the curve. For example, for an orbit with an eccentricity of

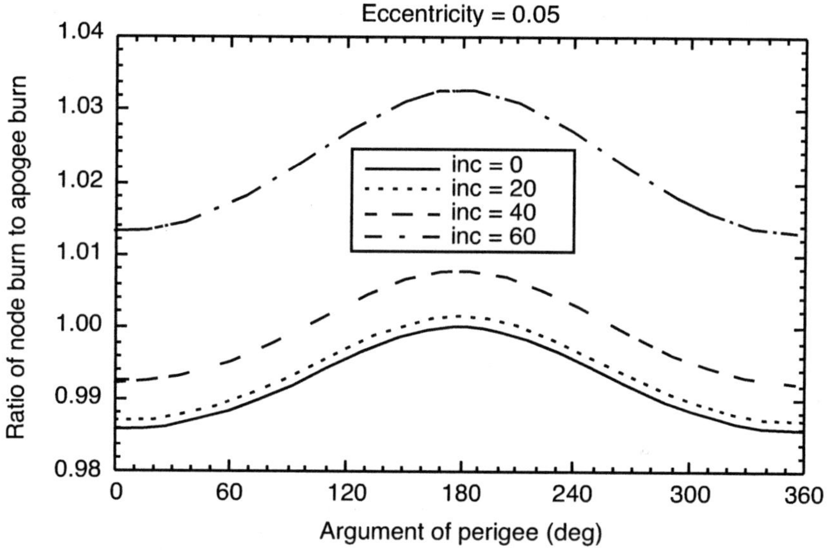


**Fig. 4.32. Ratio of burns showing relative effectiveness of node burns (e = 0.05).**

0.01 and true anomaly at node of 60 deg, the inclination must be greater than about 25 deg if the node burn is to be better. If the true anomaly at node is 120 deg for the same orbit, the inclination must be greater than approximately 15 deg. This behavior changes rapidly for increasing eccentricities. For relatively

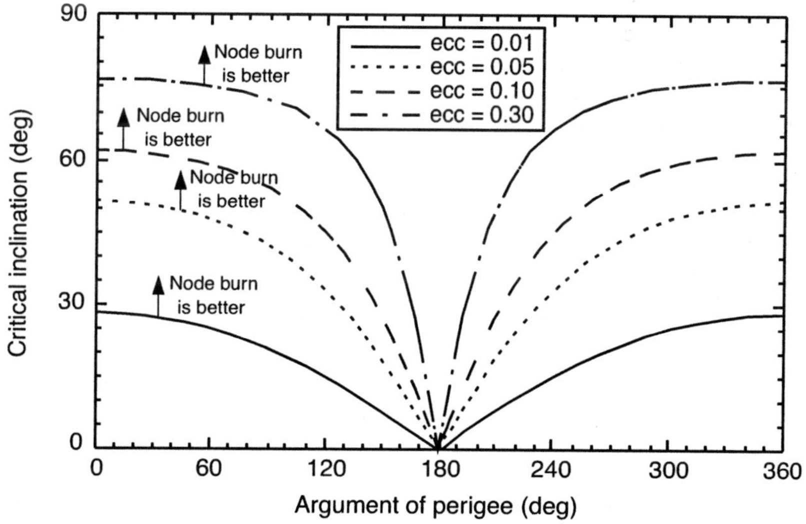


**Fig. 4.33. Critical inclination where out-of-plane node burn is better than in-plane apogee burn.**

mild eccentricities, the node burn is more efficient only for rather high inclinations or when the apogee is itself located close to the node. For example, an orbit with an eccentricity of 0.3 has to have its apogee point within 40 deg of the node if the inclination is less than 60 deg, while a Molniya-type orbit with an eccentricity of 0.7 must have its apogee within 17 deg of the node for the same inclination if the out-of-plane burn option is to be pursued (not shown). But for near-circular orbits, the out-of-plane burn is more efficient in terms of altering the orbit elements, in turn causing the largest change possible in the $J_2$ node precession for a wide range of arguments of perigee and inclinations. However, for near-equatorial orbits, the eccentricity has to be nearly perfectly circular for this to be the case.

Note from Fig. 4.33 that for each eccentricity, a critical inclination exists where any argument of perigee can yield better maneuvers than in-plane apogee burns (i.e., where the curves intersect the left-hand axis). This can be found by setting the argument of perigee to be zero (i.e., with the node burn taking place at the worst possible location: when the periapse is lined up with the node crossing at which the burn is taking place) and solving for the inclination when the ratio of the node rates is equal to unity. The resulting critical inclination is (note how rapidly the value of the critical inclination changes for very small eccentricities):

$$\tan i_{cr} = \frac{\sqrt{28e}}{(1-e)} \tag{4.117}$$

### 4.6.4 DeltaV Requirements for Oblateness Burns

The deltaV requirements for near-circular orbits are found as follows. Rearranging Eq. (4.107) in terms of the time before the storm (which is equal to one-half of the total mission downtime assuming symmetry) and the deltaV required yields:

$$\Delta v \Delta t = \sqrt{\frac{\mu}{a}}\left\{\frac{\Delta\Omega}{\dot{\Omega}_{preburn}[(\mp 1)_{\phi} 7\cos\theta_{opt} - \tan i \sin\theta_{opt}]}\right\} \tag{4.118}$$

where $\theta_{opt}$ is determined from the optimal condition of Eq. (4.112). Substituting for the oblateness precession (Eq. [4.89]) yields (where $\Delta\Omega$ is in radians, $\Delta t$ is in seconds, and $\Delta v$ is in km/sec):

$$\Delta v \Delta t = \frac{(\pm 1)_{\phi} 2a^3 \Delta\Omega}{21 J_2 R_e^2} \frac{\cos\theta_{opt}}{\cos i} \tag{4.119}$$

This then represents a relation describing a specific orbit (the eccentricity is assumed to be near zero), the size of the node angle change, the magnitude of the burn, and the time before the storm at which the burn is applied ($\theta_{opt}$ is solely a function of the inclination).

Figures 4.34–4.37 show the deltaV versus time-before-storm graphs for Eq. (4.119). In representing an equation with five variables and five unknowns, several examples of simplification are used. The first is to assume a node change angle of 10 deg. In this way, it is a simple matter to scale the results linearly for

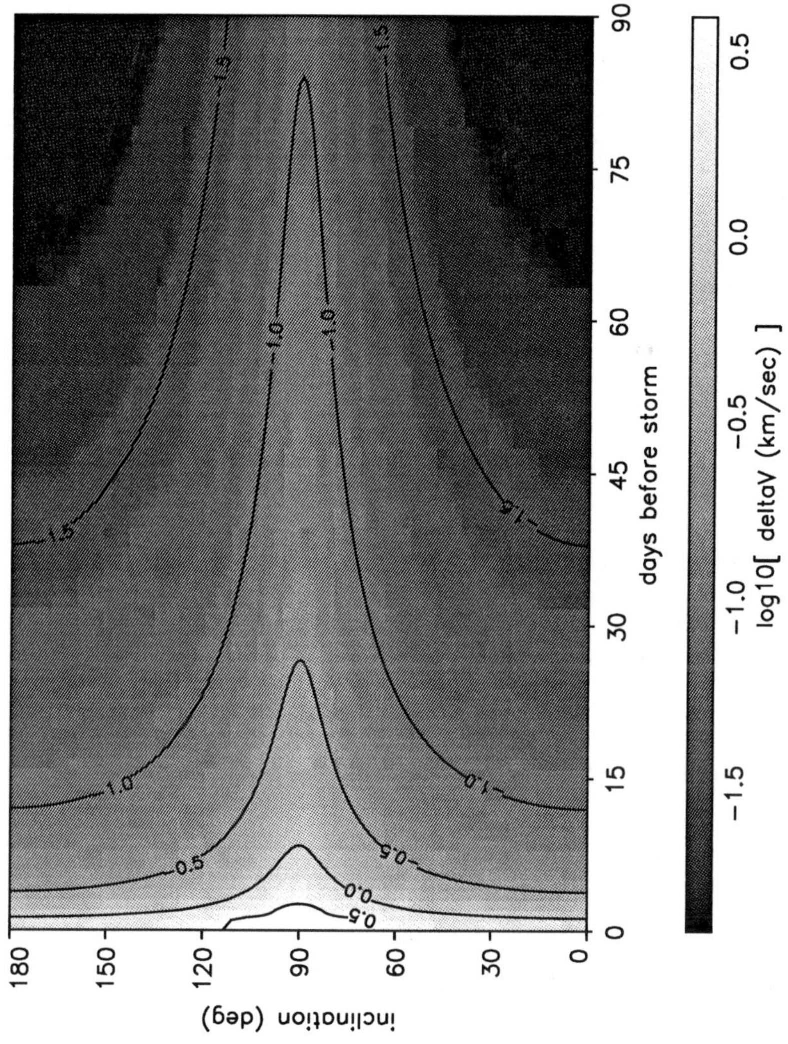


**Fig. 4.34. DeltaV requirements for 6500 km orbit (122 km altitude).**

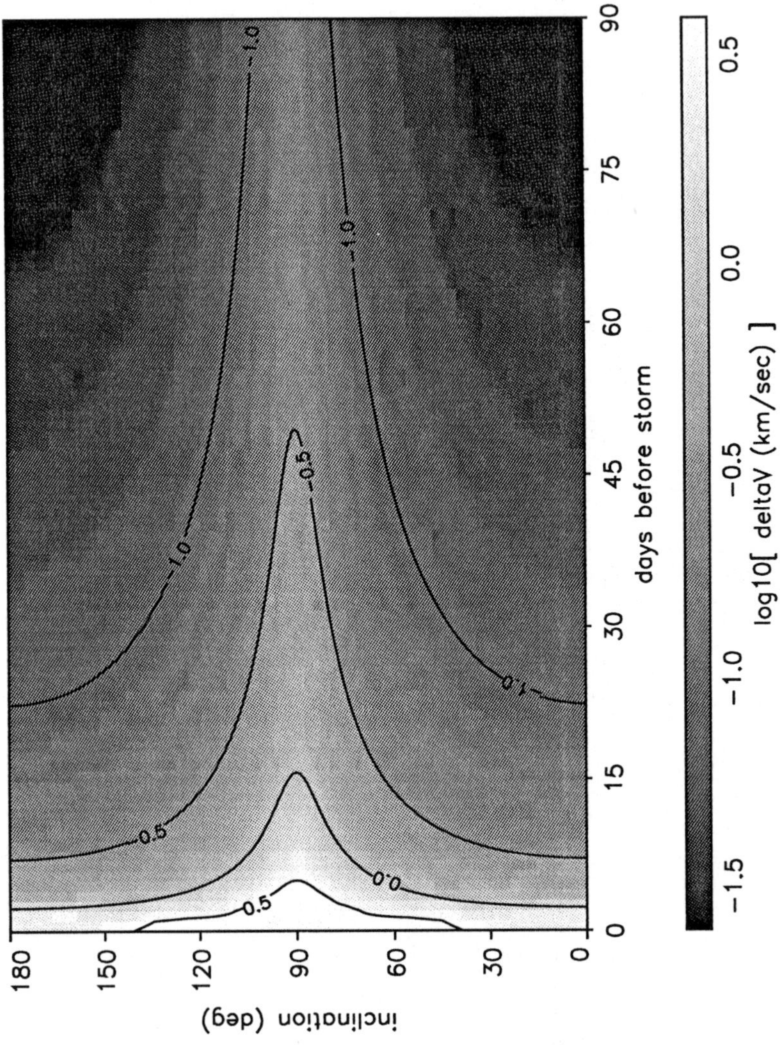


**Fig. 4.35. DeltaV requirements for 8000 km orbit (1622 km altitude).**

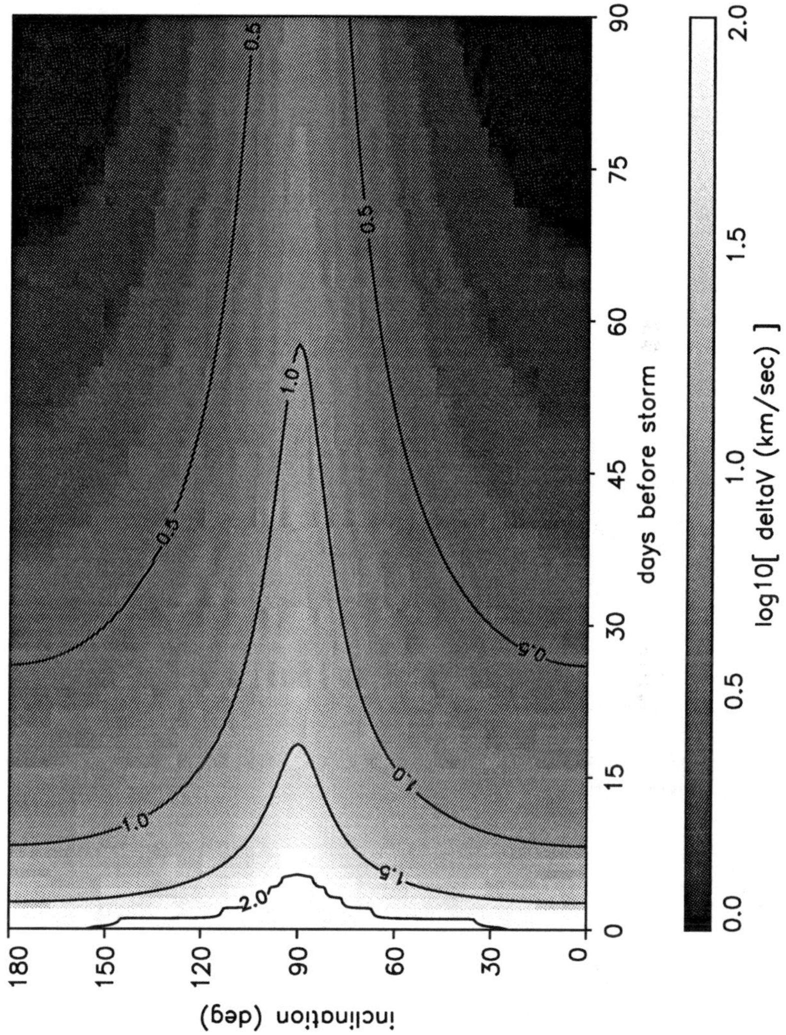


**Fig. 4.36. DeltaV requirements for 26,600 km orbit (GPS altitude).**

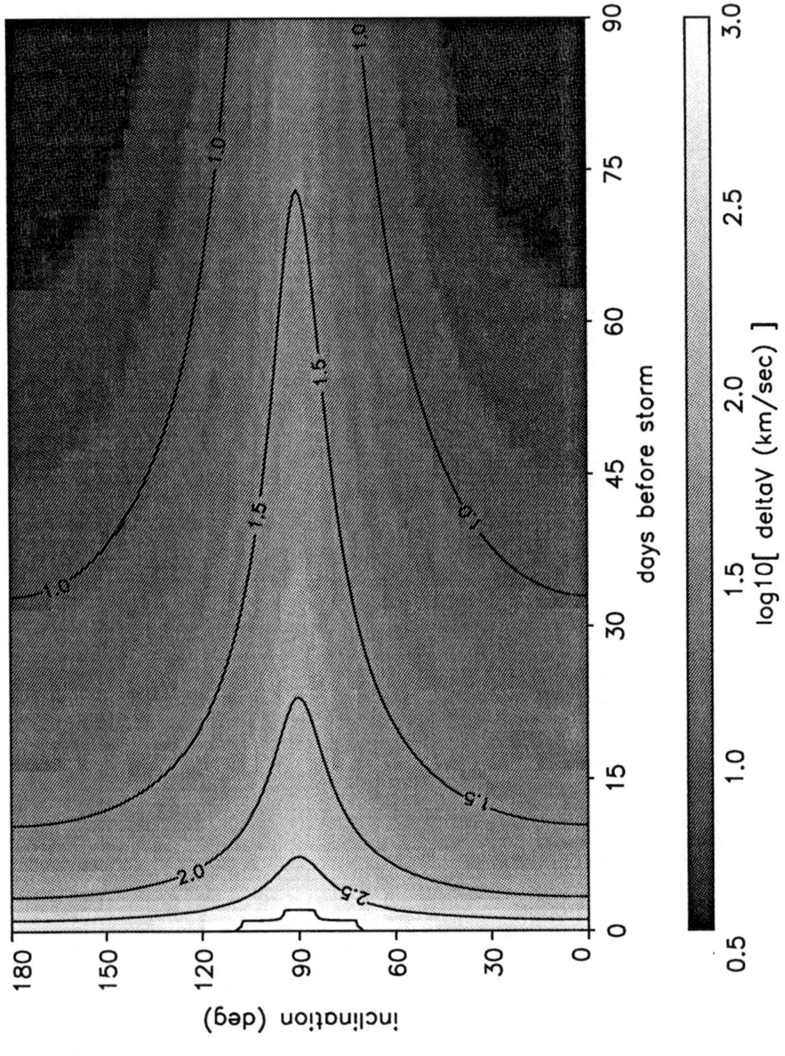


**Fig. 4.37. DeltaV requirements for 42,240 km orbit (GEO altitude).**

the desired node change angle. The second is to generate plots for specific values of the semimajor axis. Interpolation can then be used for the intermediate semimajor axis. Additionally, the absolute value of Eq. (4.119) is taken because the maneuver is symmetric about the direction of the burn $(\pm)_\phi$. Finally, the log10 of the deltaV is shown on the graphs in order to make the scale more readable. A value of 1.0 represents 10 km/sec, 0.0 corresponds to 1 km/sec, and −1.0 equates to 100 m/sec. Note again that these values are for a 10-deg change in the node angle. If only a 1-deg change is desired, the values should be divided by 10 (*not* the plotted log values).

From Figs. 4.34 to 4.37, it can be seen that the general behavior is the same for all orbit altitudes. As the time before storm increases, the smaller the deltaV that is required to effect the desired change. In essence, as the time increases, a smaller change in the node rate precession is necessary in order to accomplish the desired drift to the eclipsed node angle.

There is an additional inclination dependence. As the inclination approaches polar values, the deltaVs become very large. This is a result of the change in value of the Earth oblateness precession (Eq. [4.89]). As the inclination increases, the precession decreases and eventually reaches zero for an exact polar orbit. In the near-polar case, a substitution can be made in terms of the co-inclination ($i' = 90 - i$) and the co-theta ($\theta' = 90 - \theta$) with Eq. (4.119) being rewritten in terms of small-angle formulas in $i'$ and $\theta'$. The inclination dependence will drop out, and what is left is the maximum limiting value of the deltaV as a function of orbit height and desired node change angle:

$$\Delta v \Delta t = \frac{(\pm 1)_\phi 2a^3 \Delta\Omega}{7J_2 R_e^2} \tag{4.120}$$

The deltaV values for any of the orbits are rather large. For a 15-day time-before-storm maneuver, a 10-deg node angle change and a 60-deg inclination, the deltaVs required for the given LEO orbits are 155, 194, 239, and 290 m/sec for semimajor axis of 6500, 7000, 7500, and 8000 km, respectively. These values are rather high and may be prohibitive in terms of deltaV usage. If a satellite faces an extreme risk from a meteor storm or if the satellite has not yet entered its operational phase (i.e., spares), it may be practical to conduct the burn far in advance of the storm in order to reduce the fuel consumption to an acceptable level.

While Figs. 4.34 and 4.35 show various LEO altitude satellites, Figs. 4.36 and 4.37 show deltaV requirements for MEO (GPS) and GEO altitude orbits. For these orbits, the deltaVs are noticeably larger than for the LEO cases (deltaV is the cube of the semimajor axis). Even for a maneuver performed 90 days before the storm, the deltaVs required for a near-equatorial orbit (lowest value possible) are 0.91 km/sec and 3.63 km/sec for a 10-deg node change. Moving the node is therefore even more restrictive for higher-altitude orbits than it is for LEOs.

From an operational standpoint, the additional issue of orbit recovery must be considered. Specifically, if the orbit is changed from a desired node orientation under an operational situation to the optimal eclipsing node, it is logical to assume that it will need to be changed back once the threat is over. Therefore, the total deltaV to safe the satellite to the eclipsed position and then restore it to the operational node will be larger than the value presented in Eq. (4.119) (Figs. 4.34–4.37).

The total deltaV to safe and then recover the satellite is not simply twice the deltaV to safe the vehicle into the eclipsed node, but rather approximately four times unless the new orbit is allowed to drift all the way around back to the original node. This would limit the total deltaV to just twice the values shown, but at a potentially significant increase in mission downtime. To change and then recover the original orbit without letting the drift occur all the way around, not one but four burns must be applied. The first is the just-analyzed node rate change maneuver applied at a certain time-before-storm. This places the node into the eclipsed location, which is not the same as the desired operational angle. The second burn can be applied at the time of the storm and recovers the original orbit's shape, size, and inclination; however, the node is still phased into the wrong location from an operational perspective. In order to phase the node into the proper operation angle, two more burns must be applied. The third burn can be applied at the same time as the second (i.e., as a combined maneuver), but must change the orbit elements such that the node rate change is opposite in the sense of the original change. That is, if the original node rate was sped up in order to move the node into the eclipsed location, then the third burn slows the node rate in order to let the operational node catch up to the actual. The fourth burn is applied once the actual and operational nodes match, and recovers the original orbit's complete set of parameters. For small changes in the node, these burns will be nearly symmetrical; thus, the total deltaV to perform such a maneuver is actually about four times the values for the single burn as depicted in the figures.

Using so much fuel can shorten the mission lifetime beyond practicality. In accordance with this, vehicle operations also require that the satellite be able to perform its mission without significant downtime. When the node is not in its operational location, loss of mission may be a consequence. To minimize the deltaV used, the downtime could be on the order of weeks for a significant node change. This, too, can make moving the satellite's orbit impractical.

Overall, several factors have to be considered when deciding whether changing the node is a viable option. While such a maneuver as described here is possible, significant amounts of deltaV are consumed (thus shortening the mission lifetime), significant downtime may result (thus negatively affecting the mission), and the inherent risk when activating the engines on an orbiting vehicle is faced. These factors make this node-change maneuver impractical for most meteor events except in cases of very small node changes or for nonoperational spacecraft that can apply the burn long in advance of the storm and thereby reduce the deltaV to acceptable levels (i.e., on-orbit spares). However, if a storm is especially intense and is expected to cause the effective loss of a satellite, the eclipsed node should be considered as a possible option, albeit an expensive and dangerous one, to decrease the potential threat. A summary of the procedure for nodal maneuvering is given in Algorithm 4.5.

1. Propagate the orbit to the time of the outburst, using standard techniques yielding the node at that time, $\Omega_{outburst}$.

2. Compute the optimal node as: $\sin(\Omega_{opt} - \alpha) = -\dfrac{\tan\delta}{\tan i}$

3. Desired node change is: $\Delta\Omega = \Omega_{opt} - \Omega_{outburst}$

4. Total deltaV requirements are (mission downtime for direct burns and lunar assists is assumed low):

   Direct burns: $\Delta v_{\Omega only} = 2*\left[2v_s \sin\left(\dfrac{\vartheta}{2}\right)\right]$

   where $\cos\vartheta = \cos^2 i + \sin^2 i \cos\Delta\Omega$

   Lunar assist: $\Delta v_{lunar} = 2*\left[\dfrac{1}{v_s}\left(\dfrac{\mu}{2a} - \dfrac{\mu}{R_{em}}\right)\right]$

   $J_2$ oblateness assist ($\Delta\Omega$ in radians, $\Delta t$ in days):

$$\Delta v_{J_2} = \frac{8}{86400*\Delta t_{downtime}} * \left[\frac{2a^3\Delta\Omega}{21 J_2 R_e^2}\frac{\cos\theta_{opt}}{\cos i}\right]$$

   where $\theta_{opt} = \tan^{-1}\left(\dfrac{\tan i}{7}\right)$ for near-circular orbits.

5. Portion of new orbit that gets covered: $2\zeta* = 2\sin^{-1}\left(\dfrac{R_e + h_a}{a}\right)$

6. Time in shadow: $\Delta t_{shadow} = 2\zeta*\sqrt{\dfrac{a^3}{\mu}}$

7. Benefit gained, expressed as the amount of time spent in protection:

$$R_{TM} = \frac{N_0 - N_{reduced}}{N_0} = \frac{\Delta t_{shadow}}{\Delta t_{storm}}$$

8. Applicability: LEO (inactive satellites in alterable orbits such as the Iridium spares especially); not GEO, HEO, MEO, unless timing of storm is well determined. Mission operation requirements may limit this strategy to on-orbit spares.

**Algorithm 4.5. Nodal maneuvering for time minimization benefits.**

## 4.7 Sample Case

As an example, consider a sample case consisting of a cubic spacecraft ($A_1 = A_2 = A_3$) shielded by a single wall ($\gamma_p = 2/3$ for spall production and 1/3 for charge—Sec. 4.2) with attached solar panels. This combination of physical parameters for the vehicle yields optimal pitch and yaw pointing angles for the body of 35.3 deg and 45 deg with respect to the meteor shower radiant. The solar panels are assumed to be freely rotating. The orbit plane is oriented to contain both the meteor radiant vector and the Earth-core vector. This gives the optimal attitude (body and panel) for the damage minimization and area minimization strategies and the optimal orbit orientation for the flux minimization, relative velocity minimization, and time minimization strategies. Thus the comparison will be between the best possible results from each mitigation option. The practical reality will be somewhat less than the example as depicted, but the generalized results will hold.

For the meteor outburst properties, a Leonid encounter is assumed with known mass index ($s = 2.2$), length ($t = 7200$ seconds), mean radiant of 153.45 deg in right ascension and 21.87 deg in declination, and two values of the radial component of the ZHR ($D = 296$ and 4500) representing the extremes of the anticipated range. The resulting benefits for each of the five strategies is shown in Figs. 4.38 and 4.39. The values shown consist of the benefit ratio for each strategy multiplied by 100; the figures thus show the percentages of reduction achieved through each strategy.

The area minimization scheme will provide the best results for the solar panels. Because of the geometry of the Leonid encounter, near-perfect reduction of the risk can be found if the apparent radiant is followed (Eq. [4.27]) while still maintaining over 95% of the power production requirements. If the apparent radiant is not followed as is shown in Figure 4.38, then the solar panels can still be protected almost perfectly. Shallow angle impacts do have a tendency to do appreciable damage to the surface of exposed portions of a spacecraft (Ben-Shmuel, 1999), but that is more than offset by the reduction achievable through the optimal pointing.

The optimal orientation for the body attitude pointing gives approximately a 90% reduction in the potential for damage due to charge and over 50% for spall. It must be emphasized that these values are for this sample case only. In general, the actual decrease will depend heavily upon the mass and velocity indices. But a significant reduction in the risk is possible through adoption of the optimal body attitude pointing. It must also be restated that this represents the reduction in potential damage as compared with presenting the minimum cross-sectional area to the meteor stream. Thus, the optimal body attitude represents approximately an order of magnitude improvement in limiting the ESD damage and 50% reduction in spall over the body attitude which presents the minimum cross-sectional area of the spacecraft body to the meteor stream.

The time minimization strategy provides an apparently significant amount of reduction for higher altitude orbits, but this can be misleading. In order to achieve the maximum amount for the higher orbits, precise knowledge of the timing of the storm is required. In actuality the timing will not be determinable to the necessary extent. This strategy is still best for LEO altitude orbits wherein the orbit period is approximately the same length as the storm plus its associated uncertainties. Regardless of when the storm occurs, a LEO satellite will experience a certain percentage of its orbit in shadow.

The relative velocity minimization provides apparently good results for low altitude orbits. However, the length of any meteor outburst plus the associated uncertainties will be about the same length as an orbital period for LEO. Therefore, any benefit gained will average out for these satellites. For the relative velocity minimization to work, the spacecraft has to have a near-constant velocity

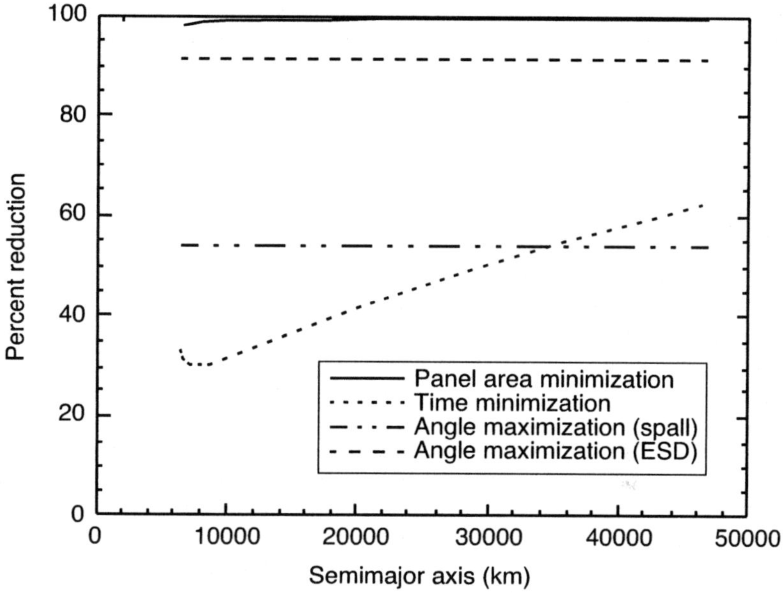


**Fig. 4.38. Risk reduction for various strategies I.**

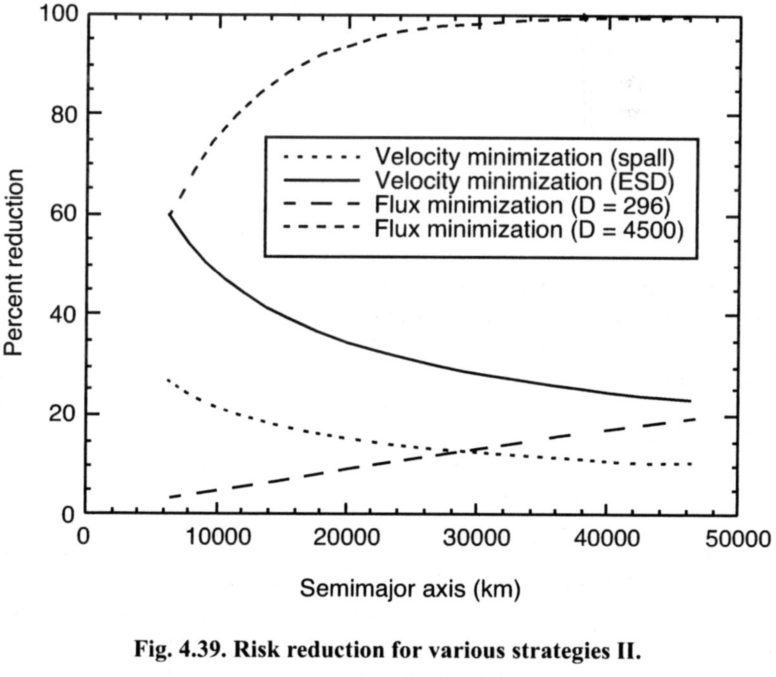


**Fig. 4.39. Risk reduction for various strategies II.**

over the course of the storm and uncertainties. Thus, relative velocity minimization is best for higher altitude orbits even though the actual benefit is less due to the slower velocities.

The flux minimization scheme is best for higher altitude orbits. At lower altitudes, the semimajor axis is not large enough to create a large difference between being on one side of the orbit versus the other. More importantly, LEO altitude orbits have an orbital period that is on the order of the length of any expected storm. As a consequence, any benefit gained through minimizing the flux will average out to a large extent over the course of a LEO orbit. This averaging is identical to the effect for relative velocity minimization. Both strategies are largely ineffective for LEOs and are better for higher altitude orbits.

It must be noted that the relative velocity minimization and flux minimization strategies can either add together or detract from each other depending upon the location of the meteor radiant, the core of the stream, and whether the orbit is prograde or retrograde. For the Leonids, if the meteor stream or streamlet core is outside of the Earth's orbit, then the benefits will add. A prograde satellite close to the core will be moving into the stream thus maximizing both the relative velocity and the encountered flux. When it is away from the core it will be moving with the meteor stream particles, minimizing the flux and the relative velocity. If the core is inside the Earth's orbit, then the benefits will at least partially cancel. For retrograde orbits, the opposite logic will prevail. If the core is inside the Earth's orbit, the benefits will add; if it is outside of the Earth's orbit, the benefits partially cancel.

## 4.8 Summary of Mitigation Options

Several dynamic strategies can be employed to mitigate the threat posed by a meteor outburst such as the Leonids of 1998–2000. With the exception of delaying the launch, none of them completely protects a satellite, but they can at least partially decrease the threat. The attitude control strategies are most applicable for active vehicles; either the damage or the chance of an impact can be reduced at very little cost in terms of fuel or mission downtime. The orbit control strategies are best in extreme meteor circumstances or for on-orbit spares. The fuel and/or mission downtime costs for these options can be large.

1. **Launch delay**. This strategy is not viable for on-orbit vehicles but does provide complete safety. However, it must be balanced against other launch and mission constraints.
2. **Satellite safing**. By powering the vehicle down into a safe mode, the effect of electrical spikes is reduced. Automatic procedures that could inadvertently hurt the vehicle in the case of a meteor strike are also prohibited. Operators have more time to assess the damage before deciding on a proper response.
3. **Attitude control for damage minimization**. It may be more important to reduce the number of meteoroids that get past a vehicle's walls than to lessen the number of particles that hit the vehicle. Attitude control can accomplish this goal. By canting the vehicle away from the minimal cross-sectional area orientation, the actual number of particles that produce spall or ESDs can be lessened. In this scenario, the number of hits increases, but the damage is decreased by an even larger amount. The optimal orientation is dependent upon the angles

at which the particles impact each of the spacecraft body's sides through the ballistic limiting equations. While the exact form of the limiting equations is uncertain at high meteoroid velocities, the interesting result remains that this dependency drops out when a cubic spacecraft is considered. The optimal orientation for protecting the internal workings of such a cubic vehicle is actually to maximize the cross-sectional area. This strategy is not applicable for exposed elements of the spacecraft such as the solar panels or antennas.

4. **Attitude control for area minimization**. The cross-sectional area of the spacecraft that is exposed to the particles is reduced through attitude orientation. This reduces the probability of an impact occurring and can potentially be conducted separately for the vehicle body, the sensors, or the solar panels. Attitude control is also the cheapest dynamical strategy in terms of fuel consumption, the safest, and the fastest in recovering the original mission once the storm has passed.
5. **Flux minimization**. By controlling the in-plane location of the satellite, the actual amount of flux faced by the spacecraft can be reduced. If the vehicle is at the point in its orbit where it is closest to the meteor stream core as the Earth passes through the stream, it will experience a potentially large difference in the flux as opposed to being at the point furthest from the meteor stream center. This benefit must be balanced by the cost, risk, and mission downtime necessary when moving satellites around in their orbital plane.
6. **Relative velocity minimization.** Controlling the in-plane location can also reduce the relative velocity between the spacecraft and the particles. This in turn reduces the levels of particle penetration and plasma discharge. This strategy, like strategy 3, does not reduce the number of particles that hit, but instead decreases the potential for the damage they cause. Both of the in-plane strategies (this one and flux minimization) are applicable only for non-LEO types of orbits, where both the velocity and position of the spacecraft are fairly constant over the outburst duration, plus the associated uncertainties. These two strategies can also complement or counteract each other. Care must then be taken in applying the in-plane options in order to make sure that the actual benefit gained is worth the cost in fuel and mission downtime.
7. **Time minimization through Earth shadowing**. The node of an orbit can be moved around through various means to an optimal location, maximizing the amount of the orbit that is shadowed. LEO satellites benefit most from this strategy. Being close to the Earth, they can receive shadowing for a third or more of the orbit. For higher-altitude satellites, the timing of the storm must be known precisely for this to be a viable option. With the typical levels of uncertainties associated with outbursts, this precise timing requirement eliminates high-altitude vehicles from consideration of node control as a viable tactic. This is the most expensive strategy in terms of fuel consumption and mission downtime; it should be used only in extreme circumstances for active satellites. For inactive satellites where mission downtime is not of major concern, this can still be a useful strategy. Spare constellation satellites, such as those used in Iridium's sparing technique, can be moved far enough in advance of an anticipated storm so that the fuel consumption is reduced to an acceptable level without impacting the active mission.

Table 4.3 summarizes the strategies in terms of the cost, benefit, and applicability for each range of spacecraft. A positive benefit ratio denotes a reduction in the risk that a satellite will face, while a negative benefit ratio implies an increase in the risk. Operational considerations may further restrict the applicability of these strategies, but these issues are satellite specific. As a general example, while attitude control options would be applicable to all vehicles, the orbit control options may result in large fuel and mission downtimes and may be most appropriate for on-orbit spares.

**Table 4.3. Mitigation Strategy Cost/Benefit Overview**

| Mitigation strategy | Cost | Benefit ratio | Applicability |
|---|---|---|---|
| Safing the satellite (Sec. 4.1) | No fuel, minimal downtime | None for mechanical damage; electrical damage reduced | All satellites |
| Attitude control for spall or ESD minimization (Sec. 4.2) | Minimal fuel, minimal downtime | $R_{DM} = 1 - (\psi)^{\gamma_p' - 1}$ where $\gamma_p' = -3(1-s)\gamma_p$ and $\psi = \cos\phi_{opt}\cos\theta_{opt}$ | Satellites of regular shape, depending upon individual vehicle properties |
| Attitude control for area minimization (Sec. 4.3) | Minimal fuel, minimal downtime | $R_{AM} = 1 - \frac{A_L}{A_L(s)}$ | Satellites with solar panels or other exposed equipment |
| Orbit control for flux minimization (Sec. 4.4) | Large fuel, short downtime to small fuel, long downtime | $R_{FM} = 1 - 10^{-Da\cos\varphi}$ | GEO (large semi-major axis); less so for MEO, HEO; little for LEO; best for on-orbit spares |
| Orbit control for relative velocity minimization (Sec. 4.5) | Large fuel, short downtime to small fuel, long downtime | $R_{RV} = \gamma_p'\left(\frac{\bar{v} \cdot \bar{V}_\infty}{V_\infty^2}\right)$ where $\gamma_p' = -3(1-s)\gamma_p$ | GEO (large semi-major axis); less so for MEO, HEO; not for LEO; best for on-orbit spares |
| Orbit control for time minimization (Sec. 4.6) | Very large fuel or long mission downtime | $R_{TM} = \frac{\Delta t_{shadow}}{\Delta T_{storm}}$ | LEO (especially spares and inactives); other categories require the time of storm to be well-determined |

# 5 The Iceman Cometh: Threats of the Future

## 5.1 Generalized Threat from Meteoric Phenomena

In previous chapters, the threat of a meteoric storm such as that produced by the Leonid meteors once or twice every 33 years was examined. However, while this depicts the threat of a particular phenomenon, it does not completely describe the general threat to spacecraft from the presence of meteors. Other storms have been known to exist and could possibly strike again. In addition, the solar system is not the clean, vacuous place it was once thought to be.

As a consequence, the threat to orbiting satellites from meteoroids can be classified into four regimes.

- Sporadic meteors that inhabit space but are not associated with any known particular stream (solar system dust)
- Meteor showers that are mated with known streams and occur each year in a fairly consistent manner (Geminids, etc.)
- Meteor outbursts that generally are associated with previously known streams but reappear with nonannual periods of unpredictable intensity (1998–1999–2000 Leonids, 1985 Draconids) and other outbursts from unexpected directions with unknown origins (1937 Corvids)
- Meteor showers that are focused by the presence of the Earth's mass cause locally intense meteor activity in small regions in the antiradiant direction that can negatively impact satellites (Chapter 3)

The purpose of the first part of this chapter is to quantify the relative threat each of these four regimes poses to Earth orbiters. The absolute threat is a function of the mass flux distribution, but it must be reiterated that this computation contains high levels of uncertainty (Sec. 1.4). Even so, an approximate level of the threat can be determined. Fifty known annual showers (Jenniskens, 1994) and 62 known outbursts from the last two centuries (Jenniskens, 1995) are listed in the appendixes and will provide the basis for the examination.

The remainder of the chapter examines the different types of meteor outbursts that warrant attention, with an eye toward predicting future occurrences. The Leonids and Draconids are analyzed as the two most dangerous examples of near-comet outbursts. A similar examination is performed for the Lyrids, the Ursids, and the $\alpha$ Monoceratids as the most threatening of the far-comet outbursts. Finally, the probability of a cometary breakup occurring such as the Andromedids of the previous century is examined.

### 5.1.1 Sporadics

Sporadic meteor activity can be characterized in terms of the Earth moving about the Sun through a dust cloud of a variable distribution (Fig. 5.1). As such, the sporadic activity can be expected to consist of a constant component with an additional annual variation. However, it has been shown that the annual variation of the sporadics is very slight, is noisy, has a latitude dependence, and is influenced by observer biases (Jenniskens, 1994). The annually averaged ZHR for the sporadics taken throughout an 11-year time span as observed from 1981 through

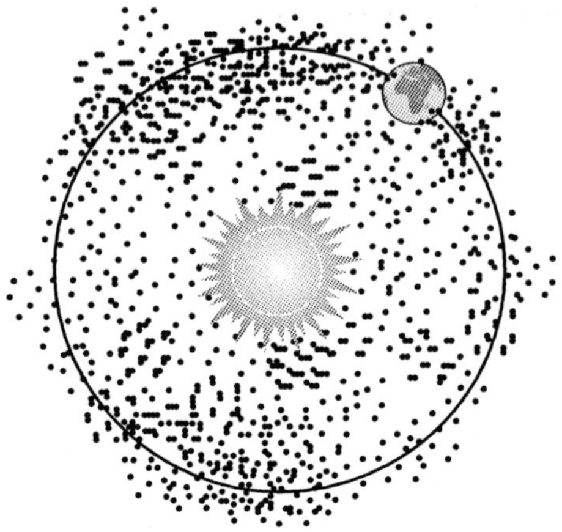

**Fig. 5.1. Annual motion of Earth through solar system dust cloud.**

1991 is ~9.0 with a mean approach velocity of ~20 km/sec (Jenniskens, 1994). The resulting cumulative flux for a dangerous mass particle (~$1.2 \times 10^{-6}$ g, Eq. [1.43]) from the full Grün model (Eq. [1.28]) is:

$$\Phi_{sporadic} = 3.98 \times 10^{-8} \text{particles/m}^2 - s \tag{5.1}$$

### 5.1.2 Annual Showers

The showers used in this study are those for which activity curves have been recently derived (Jenniskens, 1994), with the data from that study reproduced in Appendix B. The activity curve defines the level of ZHR activity as a function of time for a given meteor shower as (Sec. 1.4.1):

$$\text{ZHR} = \text{ZHR}_{max} 10^{-B|\lambda - \lambda_{max}|} \tag{5.2}$$

where $\text{ZHR}_{max}$ is the maximum zenith hourly rate for the shower under consideration, $B$ is the slope of a straight line in a log-normal plot (experimentally fitted from observational data), $\lambda$ is the solar longitude, and $\lambda_{max}$ is the solar longitude at the peak value of the ZHR.

The average ZHR over an arbitrary year is simply the sum of the averaged values for all meteor showers, where each shower average is given by:

$$\text{ZHR}_{ave} = \frac{1}{360°}\int_{\lambda_1}^{\lambda_2} \text{ZHR}_{max} 10^{-B|\lambda - \lambda_{max}|} d\lambda \tag{5.3}$$

where $\lambda_1$ and $\lambda_2$ are the beginning and ending solar longitudes of the meteor shower in question. Summing over all the meteor showers of Appendix B gives the annually averaged ZHR caused by meteor showers as ~8.4.

Compared with the sporadic ZHR of ~9, this value implies that meteoroid streaks seen by an Earth-based observer that occur do so in a ratio of about 1:1

between the annually predictable shower-produced meteoroids and the randomly encountered sporadic solar system dust cloud particles. However, as has been pointed out, the mass index of the streams is somewhat less than it is for the sporadics (Sec. 1.4) because the Poynting-Robertson drag acts to rid the stream of its smaller particles. As a consequence, the dangerous-level flux for the streams will be somewhat less than it is for the sporadics, with occasional peaks for particularly active showers. Figure 5.2 shows the dangerous-level flux over the course of a year for the known showers, along with the sporadics. Note that this is the dangerous level of flux as determined by Eq. (1.43); the slower-moving sporadics (mean of ~20 km/sec) must also have a greater mass in order to create the same visual streak than most meteor streams, according to the mass-luminosity relation of Eq. (1.39). Coupling these two facts implies that even though the average ZHRs during a year are roughly the same for the sporadics and the annual showers, the dangerous flux caused by the sporadics is about six times more prevalent than the average dangerous flux for the showers.

### 5.1.3 Outbursts

A similar analysis can be performed with the 62 meteor outbursts listed in Appendix C. Outbursts are here defined as either nonannual periodic increases in the usual annual showers (i.e., 33-year period of the Leonids or the 6.6-year period of the Draconids) or as singular shower events (i.e., Corvids of 1937). The activity curve of Eq. (5.2) can therefore be applied to outbursts as well as showers.

However, because outbursts occur at a very infrequent rate (estimated at one to two events per year), a larger time interval must be examined in order to produce

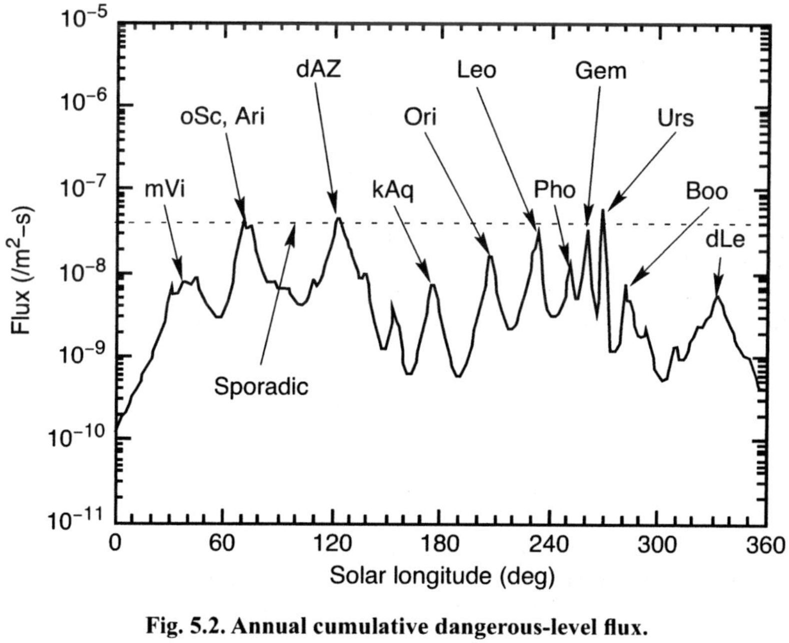


**Fig. 5.2. Annual cumulative dangerous-level flux.**

a meaningful measurement. Appendix C therefore contains data from known outbursts dated over the last two centuries (Jenniskens, 1995). The annual average ZHR over that time span is 0.111, much less than the sporadics and annual showers. However, the issue to satellites is not the average over a long time span, but rather the short-term flux that these events can produce. This is shown in Fig. 5.3 for the outbursts of Appendix C. Clearly, the dominant events are the near-comet outbursts of the Leonids and Draconids, along with the cometary breakup of the Andromedids.

In comparing Figs. 5.2 and 5.3, it can be seen that over the last 200 years, about 30 outbursts have occurred that have produced fluxes of at least an order of magnitude larger than the typical meteoroid environment. Meteor observations were somewhat spotty in the 1800s, and the levels of flux activity shown for these early outbursts involves considerable guesswork; so if only the last century is examined, fully 20 outbursts of significant levels have occurred. In other words, operators need to be aware that potentially threatening outbursts occur about once every 5 years.

### 5.1.4 Gravitational Focusing

The gravitational focusing of the Earth causes meteors to alter their path and intersect orbits along the antiradiant vector, causing potentially very intense activity in a small region (Chapter 3). For the Leonid shower, this activity does not begin until an altitude of ~256,000 km is reached and so is not a factor for Earth-orbiting satellites. However, if the meteor shower is moving more slowly than the Leonids, the initial crossover point will begin much closer to the Earth, thereby potentially intercepting the orbits of existing satellites.

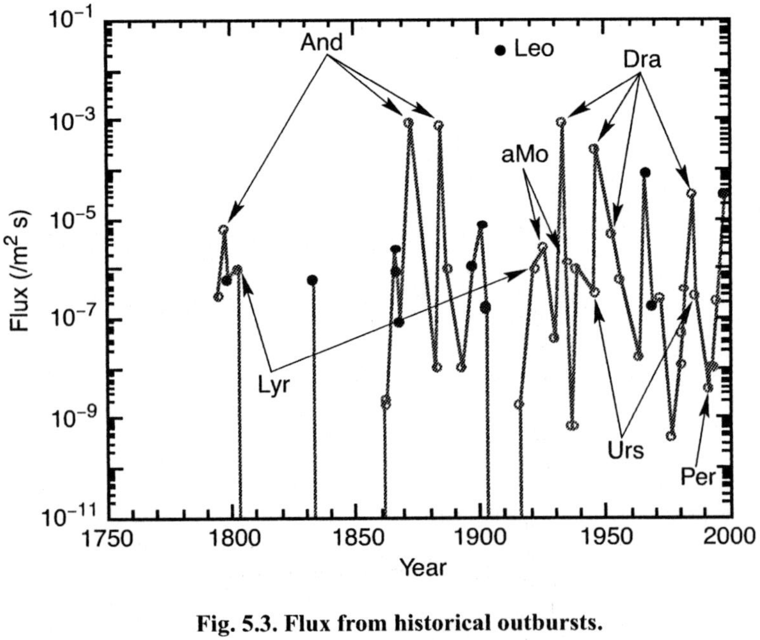


**Fig. 5.3. Flux from historical outbursts.**

In quantifying this threat, both the annual showers and outbursts will be included. Outbursts are rare events and have already been shown to be small with respect to the sporadics and the annual showers, but they can be dynamically predictable and of significant size (e.g., Leonids and Draconids). Because a methodology of examining outbursts in the same context as the showers and sporadics has already been derived in the previous sections, these outbursts also can be examined along with the annual showers without difficulty. The sporadics will not be included since their distribution is random, and while the gravitational focusing still alters their path, it will not do so in a manner that creates the intense focused region associated with a well defined meteor stream.

The probability that a satellite will be in the focused region for a particular shower is the ratio of the total area swept out by the satellite to the area of the focused region adjusted for the nodal period times the shower duration fraction in terms of the annual cycle:

$$P = \left(\frac{A_{\text{fr}}}{A_T}\right)\left(\frac{1\,\text{yr}}{TP_{\dot{\Omega}}}\right)\left(\frac{t_{\text{shower}}}{1\,\text{yr}}\right) \tag{5.4}$$

The justification for this expression is as follows. Consider a satellite that in a nodal period has an equal probability of being anywhere on a section of a sphere defined by the orbit elements (Fig. 5.4). Then the probability that the satellite will be in any particular section of that surface is the ratio of the sectional area to the total area (assuming the satellite is moving with a near-constant velocity, i.e., in a near-circular orbit). However, this does not take into account two factors—the orbit of the Earth and the shower duration. To allow for the travel of the Earth in its own orbit, the area ratio swept out in one nodal period must be scaled by the ratio of the nodal period to the annual cycle of the Earth. Similarly, the fact each shower is of lesser duration than a complete year must be included.

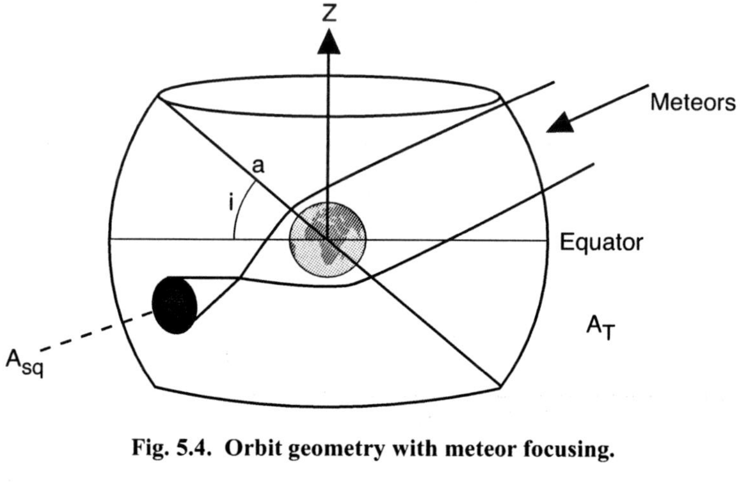


**Fig. 5.4. Orbit geometry with meteor focusing.**

The final expression describing the probability that a satellite with known orbit parameters occupies the focused region for a particular shower when the shower is occurring is given by Eq. (5.4).

If a small satellite is assumed to travel about the Earth in a near-circular orbit with known semimajor axis $a$ and inclination $i$, then the area of the shell swept out over one nodal period is given by the expression:

$$A_T = 4\pi a^2 \sin i \tag{5.5}$$

The nodal period is a function of the value of the Earth oblateness, radius, and mass constant, $J_2$, $R_e$, and $\mu$, respectively, and the same orbit parameters as before:

$$TP_{\dot{\Omega}} = \frac{4\pi a^{7/2}}{3J_2 R_e^2 \sqrt{\mu}\cos i} \tag{5.6}$$

The variable $t_{\text{shower}}$ is the temporal duration of each of the showers of interest from Appendix B. The area over which the focused region exists is simply:

$$A_{\text{fr}} = \pi x_{\text{fr}}^2 \tag{5.7}$$

where $x_{fr}$ is the perpendicular distance from the antiradiant vector (center of the focused region):

$$x_{\text{fr}} = \frac{4}{3}\frac{\sqrt{2\mu a}}{V_\infty}\Re_\phi^{-1} \tag{5.8}$$

where $\Re_\phi$ is the level of the flux ratio selected as defining the edge of the focused region. For the purposes of this report, the edges will be determined to be that focused area where the local flux is 10 times the value of the free-space flux ($\Re_\phi = 10$). Substitution into Eq. (5.4) generates the final probability expression in terms of the orbit elements, Earth physical parameters, and the meteor shower characteristics:

$$P = \frac{2}{3\pi}\frac{\mu^{3/2}J_2 R_e^2}{a^{9/2}\tan i}\frac{t_{\text{shower}}}{\Re_\phi^2 V_\infty^2} \tag{5.9}$$

The greater threat posed to a satellite from the focused region as compared with the threat caused by the free-space flux of the same shower is simply the probability multiplied by the flux ratio integrated over the size of the focused region as determined by the chosen $\Re_\phi$ (= 10), multiplied by the flux for that particular shower (Sec. 3.3):

$$\Phi_{\text{fr}-i} = P\Re_\phi(\ln x - \ln x_L + 1)\Phi_{\text{shower}-i} \tag{5.10}$$

where $x_L$ is the limiting distance relating the flux ratio along the antiradiant vector:

$$x_L = \frac{4}{3}\frac{\sqrt{2\mu a}}{V_\infty}\Re_{\phi-\text{radiant}}^{-1} \tag{5.11}$$

The threat from the focused region then is simply the individual threats summed over all the events:

$$\Phi_{\mathrm{fr}} = \Sigma\Phi_{\mathrm{fr}-i} \tag{5.12}$$

Figure 5.5 shows the threat owing to the focused regions as a function of the inclination of the orbit for two typical altitudes: MEO (GPS) and GEO. The threat is probabilistic in nature and does not represent the actual flux if a satellite were to go through one of the focused regions. Instead, Fig. 5.5 shows a combination of the actual flux multiplied by the probability of being in the focused region per year and so represents a statistical measure of the danger. It must be noted that only those showers that will create a focused region for the given orbit are to be included in the summation. That is, the minimum crossover distance for the shower must be smaller than the semimajor axis of the orbit, and the declination of the shower must be less than the inclination. If these conditions are not met, the generated focused region will occur outside of the boundaries of the swept-out shell of Fig. 5.4, and will therefore not impact the satellite's orbit.

The dominant outbursts as well as the annual showers are included in Fig. 5.5. The Leonids do not show because they are moving so quickly that their focused region is too far out to be a threat to Earth orbiters. The focused Draconids, however, can produce a threat to spacecraft but even at its worst is still orders of magnitude lower than the unfocused event (compare with Fig. 5.2). Therefore, the danger of a satellite randomly going through one of the focused regions over the course of an annual cycle is minimal.

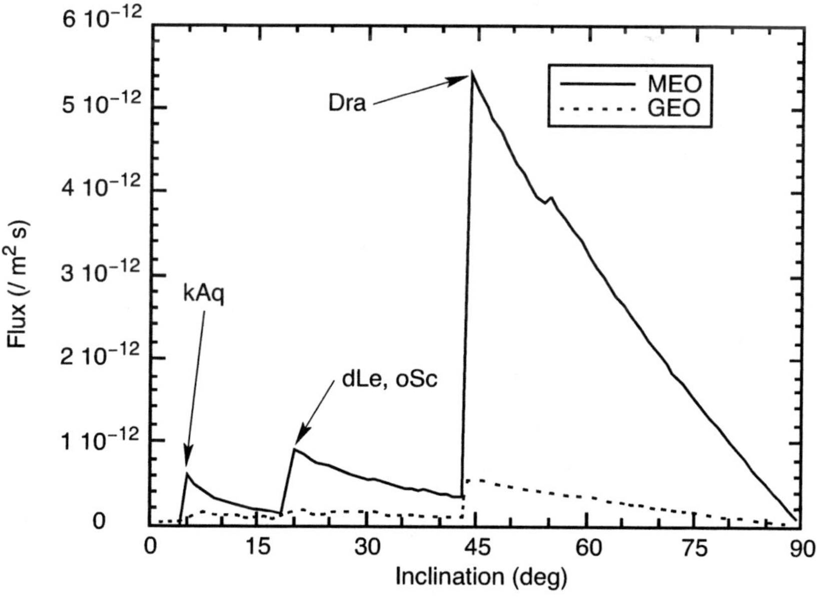


**Fig. 5.5. Probabilistic flux caused by focusing of meteor events.**

## 5.2 Near-Comet Outbursts

Outbursts from comets comprise three types of phenomena—near-comet, far-comet, and cometary breakup. Near-comet outbursts arise from meteoric material that has not yet dispersed throughout the stream but instead is still concentrated close to the parent body (Leonids, Sec. 1.2). Far-comet outbursts are caused by orbital perturbations that change the behavior of the stream, causing it to wobble or resonate in heliocentric space. That portion of the stream that the Earth goes through then changes, with the result that the Earth on different passages encounters a greater or lesser concentration of particles. The parent body in this case is far from the location in its orbit where the plane of the ecliptic is crossed (Lyrids and α Monocerotids). Finally, cometary breakup, splitting, and erratic ejection of material all can produce temporary outbursts with extremely high concentrations in meteor streams that may or may not normally produce outbursts at all (Andromedids).

### 5.2.1 Leonids

The outburst of current concern is the 1998–2000 Leonids. This has been examined extensively here in this document and in the literature at large. But because Tempel/Tuttle orbits in its cycle once every 33 years, the Leonids will become intense again sometime in the future after the 1998 event has passed. What does the future hold for this stream? Figure 5.6 shows the Leonid close encounters for the recent past, stretching over the next century. Negative distances indicate that the comet passes inside of the Earth's orbit; positive distances imply that the comet is outside. The right axis shows the number of days that the Earth crosses the point of closest approach after the comet passes that point. The 1998–2000

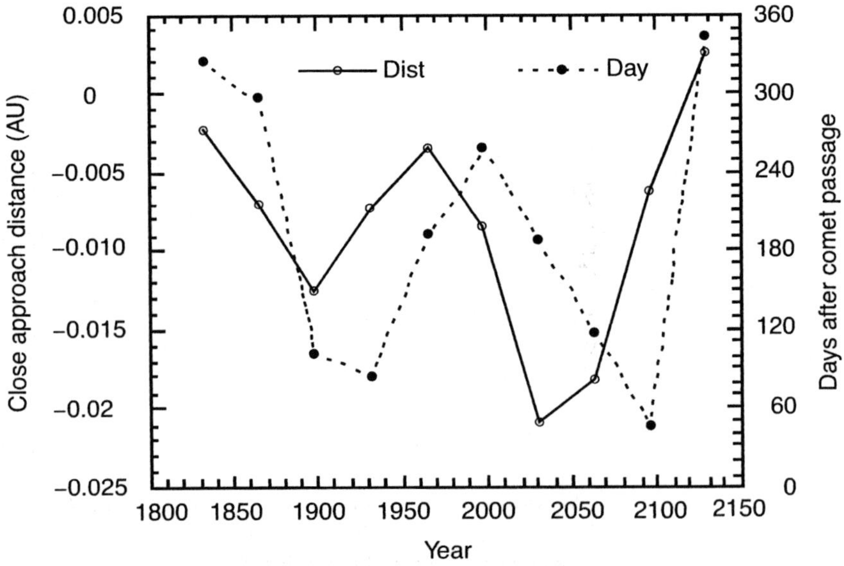


**Fig. 5.6. Leonid close encounter time history.**

events are therefore going to be the worst for the next century. By conducting a historical comparison between events, the next most significant Leonid event will be during the 2098 encounter, which will have a similar geometry to the 1998 event. The other two events in the next century will be close to 0.02 AU at their closest approach and so will present peak intensities of much less activity than in 1998–2000. After the 2098 encounter, planetary perturbations will cause the Leonids' orbit to move away from the Earth and the Leonids will cease to be a significant threat to satellite operations.

Past activity has shown that there is not a lot of correlation between the timing of the Earth and the comet for the larger streams. The Leonids are typically very active for several years around each encounter, depending upon the encounter geometry for individual streamlets generated from past perihelion passages. The Perseids, generated by the comet P109/Swift-Tuttle with a 130-year period, also exhibits this multiyear increased level of activity. Shorter-period comets do not, however, and their outbursts are typically highly dependent upon the timing of the encounter.

### 5.2.2 Draconids

One such very short-period comet produces the Draconids. The Draconids are a shower that has shown spectacular displays in the recent past. In 1933 and again in 1946, they produced ZHRs estimated to be over 10,000 and were the most intense in this century with the exception of the 1966 Leonids. Created by close encounters with the comet P021/Giacobini-Zinner, the Draconids thus present a potential problem for satellite operations that can recur with a periodicity of the parent once every 6 to 7 years. What makes matters worse, the Draconids are moving with a rather slow velocity (~23 km/sec), implying that a larger mass is necessary to create the same visual streak. In other words, a greater concatenation of particles is required to generate the same ZHR. From the meteor mass-magnitude relation of Eq. (1.39), the mass of a magnitude +6.5 Draconid is $9.3 \times 10^{-4}$ g; for the Leonids using the same relation, the magnitude +6.5 mass was $1.2 \times 10^{-5}$ g. This is nearly a two-order-of-magnitude difference, implying that the corresponding fluxes for the two events will also have a large discrepancy. This is not an insignificant result.

Consider two outburst events with the same ZHR, one for the Leonids and the other for the Draconids. Because the Draconids are moving more slowly than the Leonids, a larger mass is required to produce the same visual streak. Coupled with the flux-mass dependency of the particles, this implies that the slower-moving Draconids will have a much greater population of particles than the faster-moving Leonids. Following the analysis of Sec. 1.4, the Draconid flux for an outburst is:

$$\Phi_{\text{Dra}} = 8.48 \times 10^{-17} \text{ZHR} m^{-1.33} \tag{5.13}$$

This represents the overall flux distribution, but not the probability of impact. To find this, the minimum critical mass that has the potential to cause damage must be found from Eq. (1.43). For the Leonids, $m_c$ was $1 \times 10^{-7}$ g; for the Draconids, it is $9 \times 10^{-7}$ g. The greater overall flux of the Draconids is thus somewhat mitigated by the greater masses required to do damage to spacecraft. Figure 1.13 gave the Leonid impact probabilities for the annual shower and the worst-case

1998–2000 expected storm (ZHR ~10 000) computed over 10-min intervals.This is reproduced in Fig. 5.7 for the same Leonid storm and a Draconid outburst of the same ZHR intensity (assuming the 1946 slope values for the activity curve). Therefore, a high-intensity Draconid event has more than an order of magnitude greater chance of hurting orbiting vehicles than an equivalently sized Leonid outburst. (The "0" on the x-axis refers to the time of peak activity even though the events are months or years apart.)

With the short period of Giacobini-Zinner (6.6 years), and the greater danger, the prediction of the Draconids is important. Figure 5.8 shows the distance of closest approach and the number of days the Earth passes behind the comet. The 1933, 1939, and 1946 events all occurred with a similar close-approach distance. However, the 1939 Earth crossing was 240 days after the comet, while the 1933 and 1946 events were 93 days and 28 days, respectively. This is why those two outbursts were so large and the 1939 event was not. While the Leonids are typically active for several years around close encounter, the Draconids are not. In fact, their typical nonoutburst activity level is virtually nonexistent. A similar geometry to the great storms occurred in 1972, but the results were disappointingly slight (ZHR ~10). There is thus a noticeable uncertainty in predicting Draconid events from the geometry alone.

But if this is used as a starting criterion, two future events are potentially significant over the next 50 years. In 2018, the close-approach distance is 0.02 AU, but the Earth passes the point of closest approach only 32 days after the comet. This is similar to the 1985 encounter that produced a ZHR of ~700 (Jenniskens, 1995). The 2018 event may produce ZHRs similar to this or even more, since the 1985 close-approach distance was ~0.03 AU, or about 50% further out than

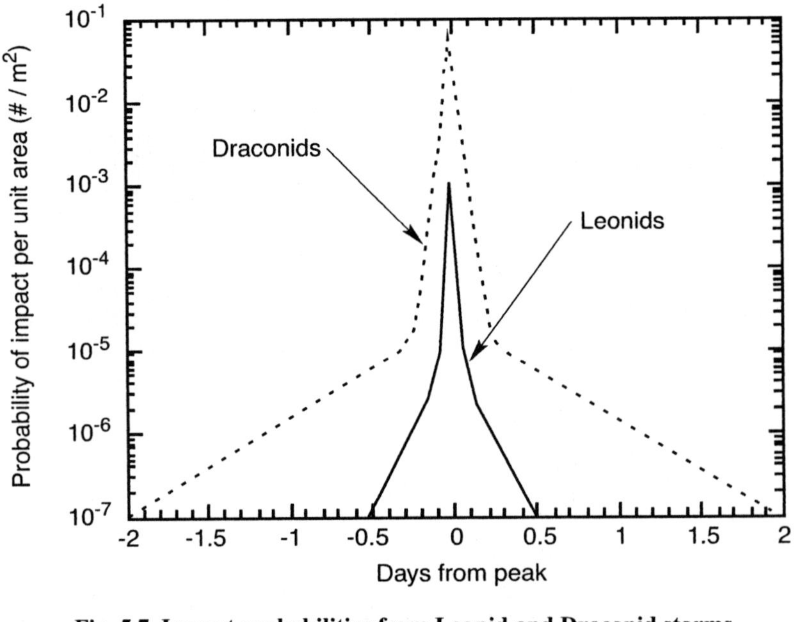


**Fig. 5.7. Impact probabilities from Leonid and Draconid storms.**

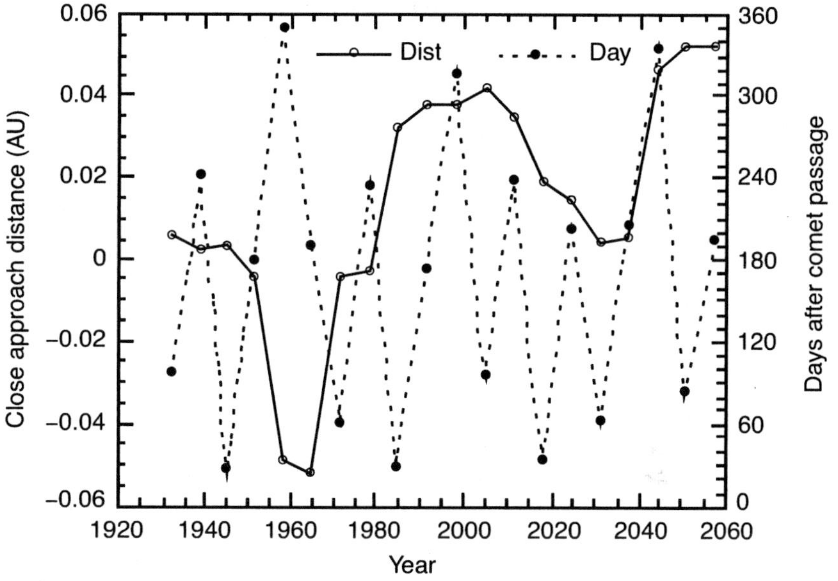


**Fig. 5.8. Draconid close encounter time history.**

the expected 2018 event. Of even more concern is the 2031 event. With a close approach distance of less than 0.004 AU and a timing of only 59 days, this will have a similar geometry to the great storms of 1933 and 1946. Table 5.1 summarizes the significant Draconid events in this century and the first 50 years of the next. It should be noted that the analysis presented here relies upon the orbit of Giacobini-Zinner as estimated during the comet's last passage in 1991. With its return in 1998–99 and the observations during that time, the ephemeris is being updated and the predicted results could change slightly. Furthermore, because of its short periodicity and potential for causing storms, comet Giacobini-Zinner should be a high priority for future observations as well.

**Table 5.1. Draconid Close Encounters**

| Year | Distance | Days | ZHR |
|---|---|---|---|
| 1933 | +0.006 | +95 | ~10,000 |
| 1946 | +0.003 | +25 | ~12,000 |
| 1952 | −0.005 | +177 | ~250 |
| 1985 | +0.032 | +27 | ~700 |
| 2018 | +0.019 | +30 | ~700? |
| 2031 | +0.004 | +59 | ~10,000? |

## 5.3 Far-Comet Outbursts

Not all the significant outbursts happen as a result of near-comet types of encounters. Far-comet outbursts occur when the parent is far from its nodal crossing and are thought to be created as a consequence of planetary perturbations causing the meteor stream to wobble and resonate in space, with the result that the Earth goes through different portions and different concentrations of the stream in different years. Because the dominant planetary perturbations are tied to Jupiter and Saturn, the wobble is largely driven by the motion of these bodies. The reflex motion of the Sun has been used to successfully predict the return of the α Monocerotids in 1995 (Jenniskens, 1997). It occurs in response to the motion of Jupiter and Saturn to maintain the solar system center of mass. Other models have been proposed (Guth, 1947; Porubcan and Stohl, 1991) but with the successful prediction of the 1995 α Monocerotids, it would appear that a wobble in the dust stream driven by planetary perturbations is a dominant mechanism in far-comet outbursts.

By restricting an examination of possible far-comet outbursts to those streams that have produced the largest displays in the past (Lyrids, Ursids, and α Monocerotids, Appendix C), Jenniskens went on to predict possible outbursts of the Lyrids in the 2016–17 and 2041–42 time frames and of the α Monocerotids in 2008 and 2018–19. The Ursids do not fit into this pattern very well and remain an unknown commodity for the future. But will these be significant for satellite operations? Using the equations of Sec. 1.4 to compute the fluxes for the three outbursts gives:

$$
\begin{aligned}
&\text{Lyrids} && \Phi_{\mathrm{Lyr}} = 4.0 \times 10^{-18} \mathrm{ZHR}\, m^{-1.16} \\
&\alpha \text{ Monocerotids} && \Phi_{\alpha\mathrm{Mo}} = 2.4 \times 10^{-18} \mathrm{ZHR}\, m^{-1.08} \qquad (5.14) \\
&\text{Ursids} && \Phi_{\mathrm{Urs}} = 2.0 \times 10^{-17} \mathrm{ZHR}\, m^{-1.12}
\end{aligned}
$$

With historical maximum ZHRs of ~860 in 1803 for the Lyrids, ~160 for the Ursids of 1986, and ~2300 in 1925 for the α Monocerotids, the worst-case impact probabilities computed over 10-min intervals for these events are shown in Fig. 5.9. Comparing these results with Fig. 5.7 shows that the far-comet outbursts are not as likely to create the dangerous environment for spacecraft that near-comet outbursts can. The probabilities presented here are for the worst case possible and still are about an order of magnitude lower than the 1998–2000 Leonids. Even so, the results in Fig. 5.9 are above the commonly acceptable threat probability of $1 \times 10^{-6}$. This type of event cannot therefore be ignored with impunity, but it is not likely to present the threat that the Leonids or especially the Draconids can.

## 5.4 Cometary Breakup

From Fig. 5.3, it can be seen that the greatest flux threat has historically come from the Leonids, the Draconids, and the Andromedids. The former two are examples of an outburst resulting from a near-comet encounter. This applies to

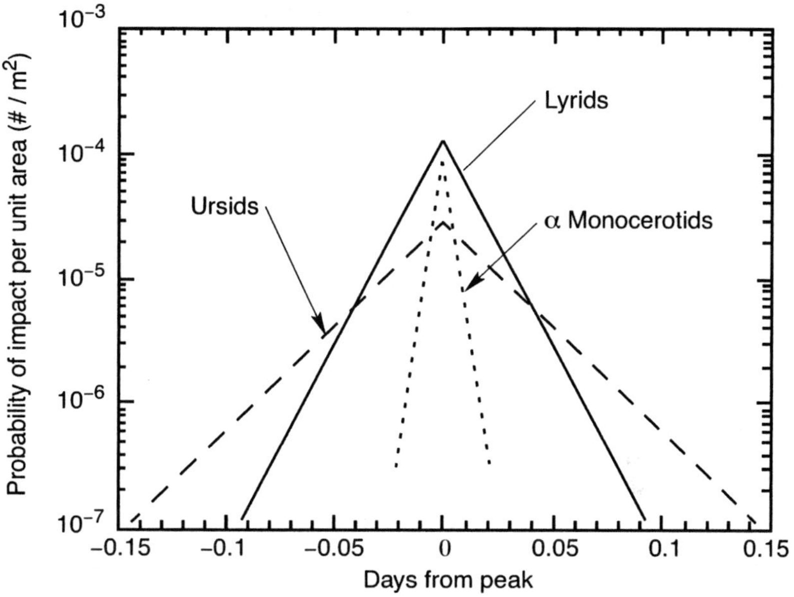


**Fig. 5.9. Impact probabilities for Lyrids, Ursids, and α Monocerotids**

the Andromedids as well, but an additional factor made this stream more intense: the breakup of its parent comet P003/Biela. This body was observed in 1772 and onward as a normal representative of comets; then when it was recovered in 1846 and 1852, it was discovered that the nucleus had split into two sections. The comet has not been seen since 1852. The subsequent Andromedid showers of 1872 and 1885 were exceptionally intense. It is assumed that Biela completely disintegrated sometime after the 1852 sighting. Using the data from Appendix C for the Andromedid storm of 1885, the expected flux is:

$$\Phi_{And} = 1.7 \times 10^{-16} \mathrm{ZHR} m^{-1.39} \tag{5.15}$$

The corresponding impact probabilities over 10-min intervals are shown in Fig. 5.10, along with the impact probabilities of the Draconid and Leonid storms from Fig. 5.7. Comparing the Andromedids to these near-comet outbursts indicates just how potentially dangerous a disintegration can be. It must be noted that P003/Biela would have had a close approach to within 0.0005 AU during the two postdisintegration encounters. However, given that the Andromedids had a noticeable outburst only once before disintegration in 1798 from a close encounter, it would seem reasonable to conclude that the large activity seen during the postdisintegration events was predominantly a result of the disintegration.

Therefore disintegration may or may not be an important phenomenon. Many comets that have been sighted in the past have been lost and not recovered, but the exact reasons for their loss are largely unknown. Comets can go through periods of quiet behavior and then suddenly reactivate. The recent identification of

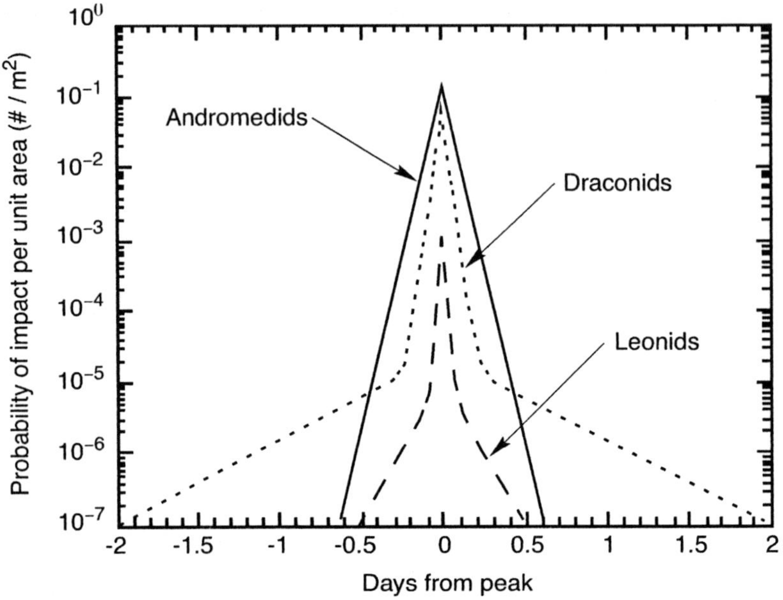


**Fig. 5.10. Andromedid impact probabilities.**

small planetoid objects in comet-like orbits (Kresak, 1991) implies that quiet comets may indeed not be very unusual. Comets are also usually seen under very favorable observing conditions, and if the conditions are not favorable on subsequent passes, recovery becomes naturally difficult. The only confidently identified example of disintegration of an Earth-crossing comet is therefore Biela. (The breakup of Shoemaker-Levy 9 was also observed, but it was never an Earth-crossing comet.) Even though complete disintegration is thought to be a rare phenomenon, comets do not behave in a homogenous manner throughout their lives.

Comet splitting is one fairly common happening. Kresak (1991) gives a statistical average of a split occurring about once every 80 revolutions for short-period comets, and once every 12 orbital periods for long-period comets. In other words, comet P055/Tempel-Tuttle, parent of the Leonids, will split about once per 2 millennia; the Draconids' parent, P021/Giacobini-Zinner, will split once in 500 years.

At infrequent intervals, erratic ejection of material can occur in the comet's behavior. (This is commonly called cometary outbursts in the literature, but referred to here as erratic ejection in order to avoid confusion with the meteoric outbursts.) This ejection is generated by unknown phenomena, although it has been put forth that a satellite of the comet periodically impacts the body (Whipple, 1983). Another possible reason is that pockets of volatiles on the comet may become active, yielding high levels of short-lived activity. The result is that temporary productions of meteoroid particles can exceed an order of magnitude or more of the entire output over a comet's period. But other than noting that erratic ejections seem to occur in pairs separated by a month or two (Kresak, 1991), the understanding of erratic ejection is itself erratic at best.

The problem with all these phenomena is that they are essentially random occurrences. No solid theory exists that would enable predictions to be made that a comet will break up, split, or erratically eject material. Therefore it is imperative, given the high fluxes from the Andromedids, that observations be made of Earth-crossing comets in a continuous fashion in order to be forewarned of any dangerous circumstances.

## 5.5 Summary of Future Threats

Meteoroids that pose a potential threat to satellite operations can be classified into four regimes:

- Sporadic (background)
- Showers (annually predictable recurrences)
- Outbursts (storms, nonannual periodic behavior)
- Focused region

Sporadic meteors arise from the background dust that resides in the solar system. Meteor showers are identified as arising from decaying or extinct comets and asteroids and make up a fairly predictable annual phenomenon. Outbursts are either random periods of intense meteor activity that cannot be predicted or are associated with nonannual periodic behavior of known streams (e.g., Leonid events of 1998–2000). Focused regions are created by known showers that are focused by the Earth's gravity into small regions of intense activity.

The average relative danger posed to satellites is greatest for the sporadics. The annually recurring showers is a close second. Both impact probabilities computed over 10-min intervals are below the usual operations threat limit of $1 \times 10^{-6}$. However, meteor outbursts can approach or exceed this level quite often. Satellite operators can conservatively expect to face this type of meteoric threat about once every 5 years on average. These events are predominantly dynamically predictable, so careful analysis can assist in avoiding the threat.

For the future of near-comet outbursts, the Leonids will apparently not be an issue after the 1998–1999–2000 encounter. Approximately a century will pass before the Leonids are in another favorable geometry to create a storm-like condition. Of more concern is the Draconids. In 2018 and 2031, P021/Giacobini-Zinner will pass close enough to the Earth that favorable geometries similar to past significant Draconid events of 1933 and 1946 will be created. Especially in 2031, the Draconids are the biggest threat foreseeable for satellite operations.

Other meteor streams can also pose a significant risk for spacecraft. In recent times, far-comet types of outbursts exhibited by the Lyrids, Ursids, and α Monocerotids have all produced impact probabilities that, while not as great as the Leonids and Draconids, are in excess of the commonly used collision probability of $1 \times 10^{-6}$. With Lyrid returns predicted for 2016–17 and 2041–42 and α Monocerotids predicted for 2008 and 2018–19, these showers—and the mechanisms that drive their outbursts—must be better understood in order to assess the threat. Ursid outbursts especially must be studied because they do not fit into the pattern of planetary perturbations well and so are still fairly unpredictable at this time.

Finally, cometary breakup can cause extremely intense meteoric events, but they occur only rarely. Only one example of breakup of an Earth-crossing comet is known during the last two centuries—P003/Biela and the Andromedid storms of 1872 and 1885. Even so, such an event can be expected to occur in the future and would be catastrophic for Earth orbiters. Because the number of spacecraft and the world's dependence upon them is expected to greatly increase in the not-too-distant future, the threat from unpredicted breakup is high. Therefore, constant monitoring of comets passing close to the Earth's orbit should be a high priority for the space community.

# Appendix

## Appendix A. Orbit Elements of Significant Comets

Tables A.1–A.3 provide the orbit elements of some of the comets associated with the most well-known storms. They are taken from the NASA Jet Propulsion Laboratory Solar System Dynamics ephemeris generator web site: <ssd.jpl.nasa.gov/cgi-bin/eph>. D003/Biela's elements are only approximate and come from a Marsden and Sekanina computed orbit as to where the comet would be in 1971 if it had not broken up.

**Table A.1. P001/Halley, P002/Encke, D003/Biela, P008/Tuttle**

| Element | P001/Halley | P002/Encke | D003/Biela | P008/Tuttle |
|---|---|---|---|---|
| q | 0.587103582 | 0.331395996 | 0.825 | 0.997733853 |
| e | 0.967276875 | 0.85001324 | 0.767 | 0.824087941 |
| i | 162.2421644 | 11.9293891 | 7.6 | 54.6925567 |
| Ω | 58.8601271 | 334.7209099 | 212.8 | 270.5487242 |
| ω | 111.8656638 | 186.2725902 | 255.1 | 206.7026769 |
| $t_{peri}$ | 2446470<br>0.9589491 | 2450592<br>0.0986444 | ? | 2449528<br>0.7895265 |
| $A_1$ | 3.88e-10 | –7.3e-11 | – | 6.31e-10 |
| $A_2$ | 1.55e-10 | –1.4e-11 | – | 1.18e-10 |
| Epoch | 2446480.5 | 2450600.5 | 1971 | 2449520.5 |
| Meteor event | Orionids, delta Aquarids | Taurids | Andromedids | Ursids |

**Table A.2. P021/Giacobini-Zinner, P026/Grigg-Skjellerup, P055/Tempel-Tuttle, P073/Schwassmann-Wachmann 3**

| Element | P021/Giacobini-Zinner | P026/Grigg-Skjellerup | P055/Tempel-Tuttle | P073/Schwassmann-Wachmann 3 |
|---|---|---|---|---|
| q | 1.033712521 | 0.996810833 | 0.97658525 | 0.932774466 |
| e | 0.706482686 | 0.663805421 | 0.90550380 | 0.694824741 |
| i | 31.8587776 | 21.0867667 | 162.4861496 | 11.4235714 |
| Ω | 195.3984211 | 213.3091870 | 235.2586540 | 69.9477887 |
| ω | 172.54333026 | 359.3316281 | 172.4971303 | 198.7679742 |
| $t_{peri}$ | 2451138<br>0.8185285 | 2450690<br>0.8034473 | 2450872<br>0.5966548 | 2449983<br>0.3894717 |
| $A_1$ | 2.35e-9 | 4.7e-11 | 5.4e-10 | 1.0697e-8 |
| $A_2$ | −5.939e-10 | −1.1e-11 | 9.129e-11 | 4.84e-10 |
| Epoch | 2451120.5 | 2450680.5 | 2450880.5 | 2450000.5 |
| Meteor event | Draconids | π Puppids | Leonids | τ Herculids |

**Table A.3. P103/Pons-Winnecke, P109/Swift-Tuttle, 1566/Icarus, 3200/Phaethon**

| Element | P103/Pons-Winnecke | P109/Swift-Tuttle | 1566/Icarus | 3200/Phaethon |
|---|---|---|---|---|
| q | 1.031721810 | 0.958217069 | 0.186611716 | 0.139791824 |
| e | 0.700374166 | 0.963589253 | 0.826864110 | 0.890041873 |
| I | 13.6188792 | 113.4266448 | 22.8702921 | 22.1042963 |
| Ω | 219.9545672 | 139.4443659 | 88.1308081 | 265.5731891 |
| ω | 180.7219863 | 153.0014540 | 31.2516141 | 321.8310975 |
| $t_{peri}$ | 2450804<br>0.5172531 | 2448968<br>0.8240133 | 2450606<br>0.6642984 | 2450814<br>0.0537185 |
| $A_1$ | 1.786e-9 | – | – | – |
| $A_2$ | 4.23e-10 | – | – | – |
| Epoch | 2450800.5 | 2448960.5 | 2450800.5 | 2451000.5 |
| Meteor event | ι Draconids (June Bootids) | Perseids | daytime Arietids | Geminids |

## Appendix B. Catalog of Known Meteor Showers

This catalog is derived from Jenniskens (1994) and Cook (1973). Note that single peak estimates for the peak + background showers are also listed. The peak + background data were used in the study, but the other values are included here for completeness. The parameters are:

- $\alpha$, $\delta$: right ascension and declination of radiant
- $V_\infty$: approach velocity in km/sec
- $\lambda_1$, $\lambda_{max}$, $\lambda_2$: solar longitude at start, maximum, and end of shower
- $ZHR_{max}$: maximum zenith hourly rate
- B: slope of log-normal estimated plot of ZHR versus solar longitude
- p, b+, b–: symmetric peak (except for Geminids), rising background, descending background ZHR and B values
- r: population index (assumed equal to $\chi$)

**Table B.1. Peak + Background Activity Curves**

| Event | $\alpha$ | $\delta$ | $V_\infty$ | $\lambda_1$ | $\lambda_{max}$ | $\lambda_2$ | ZHR | B | r |
|---|---|---|---|---|---|---|---|---|---|
| Boo-p | 232 | 45 | 43 | 278 | 282.6 | 286 | 110.0 | 2.500 | 2.5 |
| b+ | | | | | | | 20.0 | 0.370 | |
| b– | | | | | | | 20.0 | 0.450 | |
| PAu-p | 337 | 33 | 42 | 105 | 123.7 | 135 | 2.0 | 0.400 | 3.2 |
| b+ | | | | | | | 0.9 | 0.030 | |
| b– | | | | | | | 0.9 | 0.100 | |
| Per-p | 46 | 58 | 61 | 105 | 139.4 | 155 | 70.0 | 0.350 | 2.5 |
| b+ | | | | | | | 23.0 | 0.050 | |
| b– | | | | | | | 23.0 | 0.090 | |
| Leo-p | 153 | 22 | 71 | 210 | 234.4 | 240 | 19.0 | 0.550 | 3.4 |
| b+ | | | | | | | 4.0 | 0.025 | |
| b– | | | | | | | 4.0 | 0.150 | |
| Gem-p+ | 113 | 32 | 36 | 249 | 261.4 | 266 | 74.0 | 0.590 | 2.6 |
| p– | | | | | | | 74.0 | 0.810 | |
| b+ | | | | | | | 18.0 | 0.090 | |
| b– | | | | | | | 18.0 | 0.310 | |
| Urs-p | 223 | 78 | 35 | 262 | 270.3 | 276 | 10.0 | 0.900 | 3.4 |
| b+ | | | | | | | 2.0 | 0.080 | |
| b– | | | | | | | 2.0 | 0.200 | |

**Table B.2. No Peak Activity Curves**

| Event | $\alpha$ | $\delta$ | $V_\infty$ | $\lambda_1$ | $\lambda_{max}$ | $\lambda_2$ | ZHR | B | r |
|---|---|---|---|---|---|---|---|---|---|
| aHy | 133 | −11 | 44 | 285 | 299 | 315 | 2.0 | 0 | 2.8 |
| dVe | 131 | −52 | 35 | 305 | 317 | 330 | 1.3 | 0 | 3.0 |
| tCe | 210 | -41 | 60 | 305 | 333 | 345 | 4.5 | 0 | 2.6 |
| Vir | 177 | 6 | 26 | 320 | 339 | 400 | 1.5 | 0 | 3.0 |
| CAu | 283 | −40 | 45 | 52 | 55 | 60 | 3.0 | 0 | 3.1 |
| eEr | 52 | −15 | 59 | 145 | 152 | 170 | 2.0 | 0 | 2.8 |
| dEr | 55 | −2 | 31 | 215 | 228 | 245 | 0.9 | 0 | 2.8 |

**Table B.3. Single Peak Activity Curves**

| Event | $\alpha$ | $\delta$ | $V_\infty$ | $\lambda_1$ | $\lambda_{max}$ | $\lambda_2$ | ZHR | B | r |
|---|---|---|---|---|---|---|---|---|---|
| Boo | 232 | 45 | 43 | 278 | 282.6 | 286 | 133.0 | 1.800 | 2.5 |
| gVe | 125 | −47 | 35 | 275 | 285.0 | 295 | 2.4 | 0.120 | 3.0 |
| aCu | 192 | −63 | 50 | 285 | 294.7 | 310 | 3.0 | 0.110 | 2.9 |
| aCa | 95 | −54 | 25 | 301 | 310.5 | 321 | 2.3 | 0.160 | 2.5 |
| aCe | 209 | −58 | 57 | 310 | 318.7 | 340 | 7.3 | 0.180 | 2.3 |
| oCe | 177 | −56 | 51 | 318 | 322.7 | 330 | 2.2 | 0.150 | 2.8 |
| dLe | 160 | 19 | 23 | 320 | 334.0 | 345 | 1.1 | 0.050 | 3.0 |
| gNo | 249 | −51 | 56 | 345 | 352.3 | 360 | 5.8 | 0.190 | 2.4 |
| dPa | 308 | −63 | 60 | 350 | 370.4 | 390 | 5.3 | 0.075 | 2.6 |
| Lyr | 272 | 33 | 49 | 25 | 31.7 | 40 | 12.8 | 0.220 | 2.7 |
| mVi | 227 | −7 | 30 | 25 | 39.0 | 55 | 2.2 | 0.045 | 3.0 |
| eAq | 338 | −1 | 66 | 25 | 45.8 | 60 | 36.7 | 0.080 | 2.7 |
| Sco | 240 | −25 | 35 | 45 | 55.2 | 60 | 3.2 | 0.130 | 2.5 |
| oSc | 239 | −20 | 21 | 69 | 71.9 | 80 | 5.2 | 0.150 | 3.0 |
| Ari | 45 | 23 | 38 | 70 | 76.0 | 95 | 54.0 | 0.100 | 2.7 |
| Sag | 271 | −26 | 29 | 60 | 88.5 | 120 | 2.4 | 0.037 | 2.9 |
| Cet | 24 | −12 | 66 | 87 | 95.0 | 97 | 3.6 | 0.180 | 2.5 |
| Oph | 249 | −15 | 27 | 75 | 97.0 | 105 | 2.3 | 0.037 | 2.8 |

**Table B.3. Single Peak Activity Curves** —*Continued*

| Event | $\alpha$ | $\delta$ | $V_\infty$ | $\lambda_1$ | $\lambda_{max}$ | $\lambda_2$ | ZHR | B | r |
|---|---|---|---|---|---|---|---|---|---|
| tAq | 342 | −12 | 63 | 93 | 97.3 | 101 | 7.1 | 0.240 | 2.5 |
| uPh | 21 | −43 | 48 | 105 | 110.5 | 115 | 5.0 | 0.250 | 3.0 |
| oCy | 305 | 47 | 37 | 107 | 116.0 | 127 | 2.5 | 0.130 | 2.7 |
| Cap | 305 | −9 | 25 | 90 | 121.7 | 150 | 2.2 | 0.041 | 2.0 |
| dAN | 339 | −5 | 42 | 110 | 123.4 | 140 | 1.0 | 0.063 | 3.3 |
| PAu | 337 | −33 | 42 | 105 | 123.7 | 135 | 2.9 | 0.260 | 3.2 |
| dAZ | 339 | −17 | 43 | 109 | 124.9 | 141 | 11.4 | 0.091 | 3.3 |
| iAZ | 338 | −14 | 36 | 110 | 131.0 | 140 | 1.5 | 0.070 | 3.3 |
| Per | 46 | 58 | 61 | 105 | 139.5 | 155 | 84.0 | 0.200 | 2.5 |
| kCy | 289 | 52 | 27 | 125 | 146.0 | 165 | 2.3 | 0.069 | 2.2 |
| gDo | 61 | −50 | 41 | 150 | 155.0 | 160 | 4.8 | 0.180 | 2.8 |
| Aur | 72 | 43 | 69 | 150 | 157.5 | 165 | 9.0 | 0.190 | 2.7 |
| kAq | 338 | −5 | 19 | 170 | 176.5 | 180 | 2.7 | 0.110 | 2.8 |
| Eps | 103 | 28 | 71 | 195 | 206.0 | 215 | 2.9 | 0.082 | 3.0 |
| Ori | 95 | 16 | 67 | 195 | 207.9 | 225 | 25.0 | 0.120 | 3.1 |
| LMi | 162 | 37 | 61 | 206 | 209.0 | 212 | 1.9 | 0.140 | 2.7 |
| Tau | 48 | 18 | 30 | 190 | 222.9 | 260 | 7.3 | 0.026 | 2.3 |
| zPu | 117 | −42 | 41 | 226 | 231.5 | 236 | 3.2 | 0.130 | 3.4 |
| Leo | 153 | 22 | 71 | 210 | 234.4 | 240 | 23.0 | 0.390 | 3.4 |
| PuV | 135 | −46 | 40 | 255 | 251.0 | 275 | 4.5 | 0.034 | 2.9 |
| Pho | 18 | −58 | 18 | 250 | 251.7 | 255 | 2.8 | 0.300 | 2.8 |
| Mon | 100 | 14 | 43 | 256 | 260.2 | 264 | 2.0 | 0.250 | 3.5 |
| Gem | 113 | 32 | 36 | 249 | 261.4 | 266 | 88.0 | 0.390 | 2.6 |
| sHy | 127 | 2 | 59 | 259 | 264.8 | 272 | 2.5 | 0.100 | 3.0 |
| Urs | 223 | 78 | 35 | 262 | 270.3 | 276 | 11.8 | 0.610 | 3.4 |

## Appendix C. Catalog of Known Meteor Outbursts

This catalog of 60 meteor outbursts derives from Jenniskens (1995) and follows the same notation as described in Appendix A. It is divided by type: near-comet outbursts are associated with the parent comet returning to perihelion; far-comet types are possibly associated with a modulation with the orbit of Jupiter; possible outbursts are singular events not definitely classifiable in either of the other two categories. Where necessary in order to compute the average ZHR, estimates have been made based on the data for the same shower at a different time or a similar shower based on written descriptions. Leonid events in 1998 and 1999 represent mid range of the predicted storm intensities and follow the same type of activity curve seen in other Leonid events.

**Table C.1. Far-Comet Type**

| Event | Year | $\alpha$ | $\delta$ | $V_\infty$ | $\lambda_1$ | $\lambda_{max}$ | $\lambda_2$ | ZHR | B | r |
|---|---|---|---|---|---|---|---|---|---|---|
| Lyr | 1803 | 272 | 33 | 48 | 31.25 | 31.283 | 31.35 | 860 | 35.0 | ? |
| | 1922 | 271 | 34 | 48 | 31.25 | 31.290 | 31.35 | 800 | 35.0 | ? |
| | 1945 | 272 | 33 | 48 | 31.29 | 31.355 | 31.40 | 100 | 35.0 | ? |
| | 1982 | 272 | 33 | 48 | 31.30 | 31.371 | 31.45 | 250 | 33.0 | 2.9 |
| tAr | 1935 | 87 | 41 | 66 | 157.90 | 157.950 | 158.00 | 100 | 35.0 | 2.2 |
| | 1986 | 94 | 36 | 66 | 157.78 | 157.821 | 157.86 | 250 | 33.0 | ~1.3 |
| eEr | 1981 | 52 | -15 | 59 | 167.35 | 167.420 | 167.45 | 170 | 10.0 | 2.6 |
| Ori | 1993 | 95 | 16 | 67 | 201.00 | 203.600 | 210.00 | 25 | 0.6 | 2.0 |
| aMo | 1925 | 110 | -5 | 60 | 238.68 | 238.684 | 238.69 | 2300 | 115.0 | ? |
| | 1935 | 110 | -5 | 60 | 238.70 | 238.740 | 238.80 | 1200 | 69.0 | ~3.0 |
| | 1985 | 109 | -7 | 60 | 238.61 | 238.617 | 238.62 | 600 | 220.0 | ~2.7 |
| Urs | 1795 | 223 | 78 | 35 | 271.00 | 271.100 | 271.20 | 150 | 17.0 | ? |
| | 1945 | 223 | 78 | 35 | 270.55 | 270.627 | 270.65 | 120 | 17.0 | ? |
| | 1986 | 223 | 78 | 35 | 270.10 | 270.236 | 270.30 | 160 | 17.0 | 2.8 |

**Table C.2. Near-Comet Type**

| Event | Year | $\alpha$ | $\delta$ | $V_\infty$ | $\lambda_1$ | $\lambda_{max}$ | $\lambda_2$ | ZHR | B | r |
|---|---|---|---|---|---|---|---|---|---|---|
| Pup | 1977 | 110 | -45 | 19 | 32.85 | 32.973 | 33.05 | 180.0 | 10.00 | >1.6 |
| p | 1982 | 110 | -45 | 19 | 32.35 | 32.556 | 32.75 | 2500.0 | 8.40 | 1.9 |
| b | 1982 | 110 | -45 | 19 | 32.85 | 32.552 | 32.75 | 3.3 | .09 | 1.9 |
| iDr | 1916 | 238 | 55 | 19 | 97.20 | 97.413 | 97.60 | 300.0 | 8.00 | ~1.7 |
| Per | 1862 | 47 | 57 | 61 | 138.80 | 138.910 | 139.00 | 250.0 | 13.00 | ? |

**Table C.2. Near-Comet Type**—*Continued*

| Event | Year | $\alpha$ | $\delta$ | $V_\infty$ | $\lambda_1$ | $\lambda_{max}$ | $\lambda_2$ | ZHR | B | r |
|---|---|---|---|---|---|---|---|---|---|---|
| Per | 1863 | 47 | 57 | 61 | 138.60 | 138.940 | 139.30 | 300.0 | 9.00 | ? |
| | 1991 | 46 | 58 | 61 | 138.80 | 138.869 | 139.95 | 500.0 | 25.00 | 1.9 |
| | 1992 | 46 | 58 | 61 | 138.70 | 138.771 | 138.85 | 400.0 | 22.00 | 2.1 |
| | 1993 | 46 | 58 | 61 | 138.50 | 138.810 | 139.10 | 230.0 | 6.00 | 2.2 |
| Dra | 1933 | 262 | 55 | 23 | 196.20 | 196.302 | 196.40 | 10000.0 | 24.00 | 3.6 |
| | 1946 | 262 | 54 | 23 | 196.20 | 196.292 | 196.40 | 12000.0 | 17.00 | 3.2 |
| | 1952 | 262 | 54 | 23 | 196.20 | 196.241 | 196.40 | 250.0 | 25.00 | ? |
| p | 1985 | 262 | 55 | 23 | 194.30 | 194.565 | 194.90 | 700.0 | 13.00 | 3.4 |
| b | 1985 | 262 | 55 | 23 | 194.20 | 194.565 | 195.10 | 4.0 | 1.20 | 3.4 |
| Leo | 1799 | 149 | 22 | 71 | 231.90 | 232.100 | 232.40 | ? 5000.0 | 30.00 | ? |
| | 1833 | 150 | 22 | 71 | 232.20 | 232.450 | 232.70 | ? 5000.0 | 30.00 | ? |
| p | 1866 | 151 | 22 | 71 | 232.55 | 232.627 | 232.75 | 17000.0 | 30.00 | ~2.5 |
| b | 1866 | 151 | 22 | 71 | 232.55 | 232.625 | 232.80 | 1000.0 | 6.00 | ~2.5 |
| p | 1867 | 151 | 22 | 71 | 232.65 | 232.713 | 232.80 | 6000.0 | 30.00 | ? |
| b | 1867 | 151 | 22 | 71 | 232.55 | 232.713 | 232.80 | 350.0 | 6.00 | ? |
| | 1868 | 151 | 22 | 71 | 232.90 | 233.122 | 233.20 | 700.0 | 4.50 | ? |
| | 1898 | 152 | 22 | 71 | 232.70 | 233.460 | 234.00 | 1100.0 | 4.10 | ? |
| | 1901 | 152 | 23 | 71 | 232.60 | 233.460 | 234.20 | 7000.0 | 3.50 | 3.1 |
| | 1903 | 152 | 22 | 71 | 232.70 | 233.460 | 234.20 | 1400.0 | 3.50 | ~2.5 |
| p | 1966 | 153 | 21 | 71 | 234.20 | 234.468 | 234.70 | 15000.0 | 30.00 | 2.9 |
| b | 1966 | 153 | 21 | 71 | 234.20 | 234.468 | 234.80 | 150.0 | 6.00 | 2.9 |
| | 1969 | 153 | 21 | 71 | 234.50 | 234.567 | 234.60 | 250.0 | 30.00 | 3.0 |
| * | 1998 | 153 | 21 | 71 | 234.49 | 234.567 | 234.70 | 5000.0 | 30.00 | 3.0 |
| * | 1999 | 153 | 21 | 72 | 234.49 | 234.567 | 234.70 | 5000.0 | 30.00 | 3.0 |
| And | 1798 | 24 | 44 | 20 | 256.90 | 257.100 | 257.40 | ? 5000.0 | ~11.00 | ? |
| | 1872 | 24 | 44 | 20 | 246.90 | 247.015 | 247.20 | 7400.0 | 10.00 | ~3.6 |
| p | 1885 | 24 | 44 | 20 | 246.50 | 246.645 | 246.90 | 6400.0 | 9.50 | 3.6 |
| b+ | 1885 | 24 | 44 | 20 | 245.70 | 246.645 | 246.65 | 100.0 | 0.30 | 3.6 |
| b– | 1885 | 24 | 44 | 20 | 246.64 | 246.645 | 247.80 | 100.0 | 1.40 | 3.6 |
| Pho | 1887 | 24 | -55 | 17 | 252.00 | 252.200 | 252.50 | 50.0 | 1.90 | ~3 |
| | 1938 | 23 | -57 | 17 | 252.80 | 253.100 | 253.30 | 50.0 | 1.90 | ? |
| | 1956 | 15 | -58 | 17 | 253.10 | 253.440 | 253.60 | 50.0 | 1.90 | 2.9 |
| | 1972 | 25 | -57 | 17 | 252.10 | 252.400 | 252.60 | 20.0 | 1.90 | ? |

**Table C.3. Possible Outbursts**

| Event | Year | $\alpha$ | $\delta$ | $V_\infty$ | $\lambda_1$ | $\lambda_{max}$ | $\lambda_2$ | ZHR | B | r |
|---|---|---|---|---|---|---|---|---|---|---|
| kPa | 1986 | 275 | -67 | 25 | 113.90 | 114.130 | 114.40 | 60 | 30 | 2.2 |
| bHy | 1985 | 23 | -76 | 24 | 143.10 | 143.133 | 143.20 | 80 | 30 | 2.1 |
| mPe | 1883 | 339 | 22 | 16 | 229.90 | 229.900 | 229.95 | 100 | 15 | ? |
| | 1893 | 339 | 22 | 16 | 230.35 | 230.400 | 230.45 | 100 | 15 | ? |
| | 1952 | 339 | 22 | 16 | 229.65 | 229.700 | 229.75 | 100 | 15 | ? |
| aCe | 1980 | 209 | -58 | 57 | 318.42 | 318.484 | 318.54 | 230 | 60 | 2.2 |
| tHe | 1930 | 228 | 39 | 18 | 63.20 | 77.300 | 88.00 | 50 | 20 | ? |
| Cor | 1937 | 192 | -19 | 15 | 94.50 | 94.900 | 96.90 | 13 | 10 | 1.9 |
| xCp | 1972 | 302 | -9 | 16 | 180.00 | 189.040 | 197.00 | 8 | 20 | ~2.0 |
| gDe | 1930 | 312 | 17 | 60 | 79.70 | 79.720 | 79.74 | 200 | 60 | 2.5 |
| oOr | 1964 | 85 | 4 | 45 | 242.90 | 243.420 | 243.92 | 140 | 15 | 2.3 |

## Appendix D. 1998-9 Leonid Meteor Storm Encounter Data

Data computed by the author and David Lynch for The Aerospace Corporation Leonid web site: <www.aero.org/leonid>.

**Radiant (in LEO)** (Geocentric equatorial coordinates, ECI)

| | |
|---|---|
| Right ascension | 153.45° (10h 13 m 48 s) |
| Declination | 21.87° |

**Earth inertial velocity at time of storm (km/sec)**

| | | | |
|---|---|---|---|
| Geocentric-equatorial coordinates (ECI) | –24.9 | 15.6 | 6.76 |
| Heliocentric-ecliptic coordinates | –24.9 | 17.0 | 0.00 |

**Meteoroid inertial velocity at time of storm**

| | | | |
|---|---|---|---|
| Geocentric-equatorial (ECI) | 33.9 | –13.8 | –19.6 |
| Heliocentric-ecliptic | 33.9 | –20.5 | –12.5 |

**Relative velocity at time of storm**

| | | | |
|---|---|---|---|
| Geocentric-equatorial (ECI) | -58.8 | 29.4 | 26.3 |
| Heliocentric-ecliptic | -58.8 | 37.5 | 12.5 |

**Relative velocity magnitude** 70.8 km/sec

**Closest approach between Earth and core of meteoroid stream** 0.0085 AU

**Earth-core unit vector at time of storm**

| | | | |
|---|---|---|---|
| Geocentric-equatorial (ECI) | -0.5524 | -0.6787 | -0.4839 |
| Heliocentric-ecliptic | -0.5524 | -0.8152 | -0.1740 |

**Earth-Sun Unit Vector at Time of Storm**

| | | | |
|---|---|---|---|
| Geocentric-equatorial (ECI) | -0.5692 | -0.7543 | -0.3271 |
| Heliocentric-ecliptic | -0.5692 | -0.8222 | 0.0000 |

**Angle between Earth-Sun line and meteor radiant vector**

| | |
|---|---|
| 1998 | 87.54° |
| 1999 | 87.56° |

**Time of shower** November 13–20 every year

**Shower intensity**

| | |
|---|---|
| Usual | 10–15 ZHR (Zenith Hourly Rate) |
| 1998 | <200 ZHR |
| 1999 | <200 ZHR |

**Time of storm**

| | |
|---|---|
| 1998 | $20^h$ $38.4^m$ UT ($15^h$ $38.4^m$ EST) on 11-17-98 |
| 1999 | $03^h$ $21.6^m$ UT ($-1^h$ $38.4^m$ EST) on 11-18-99 |

| | |
|---|---|
| **Storm intensity** | |
| 1998 | 200–10000 ZHR |
| 1999 | 200–10000 ZHR |
| **Storm duration** | 1–2 hours |
| **Best viewing location** | |
| 1998 | 123.0° E longitude (western Pacific) |
| 1999 | 043.7° E longitude (eastern Europe) |

**Meteoroid composition**

Silicate and carbonaceous grains such as olivine and "pitch"

| | |
|---|---|
| **Particle mass** | |
| Magnitude +6.5 | $1.2 \times 10^{-5}$ grams |
| Critical for damage possibility | $1.0 \times 10^{-7}$ grams |
| **Particle flux** | |
| Shower | $1.70 \times 10^{-11}$ $ZHR_{shower}$ |
| Storm | $1.58 \times 10^{-10}$ $ZHR_{outburst}$ |

**Impact damage**

Direct mechanical cratering, plasma and EMP generation leading to electronic noise, sudden current and voltage spikes and software anomalies.

**Comet 55/P Tempel-Tuttle orbit elements** (Yeomans, 1998)

| | |
|---|---|
| Perihelion distance | 0.9765849 AU |
| Longitude ascending node | 35.25883° |
| Eccentricity | 0.9055036 |
| Argument of perihelion | 172.49731° |
| Inclination | 162.48614° |
| Time of perihelion: | 1998 Feb. 28.09666 |
| A1 | 5.4 ± 2.7e-10 AU/$d^2$ |
| A2 | 9.129 ± 0.015e-11 AU/$d^2$ |
| epoch | 1998 March 8.0 |

| | |
|---|---|
| **Next significant Leonid encounter** | |
| 2098 | (so look now or forever hold your peace) |

# Bibliography


American Meteor Society, Ltd. (web site) November 1997. Available: <www.amsmeteors.org> (current at May 4, 1999).

Anderson, B. J., and W. J. Cooke, Jr. "Calculation of Area Loss Due to Meteoroid Penetration." In *Proc. of First Leonid Meteoroid Storm and Satellite Threat Conference*. Manhattan Beach, CA: AIAA/Aerospace Corporation, April 1998.

Asher, D.J., M.E. Bailey, and V.V. Emel'yanenko, "Resonant Meteoroids from Comet Tempel-Tuttle in 1333: the Cause of the Unexpected Leonid outburst in 1998," *Monthly Notices of the Royal Astronomical Society*, to be published, 1999.

*Astronomical Almanac for the Year 1999*. Washington, D.C.: U.S. Government Printing Office, 1999.

Bate, R. R., D. D. Mueller, and J. E. White. *Fundamentals of Astrodynamics*. New York: Dover Publications, Inc., 1971.

Beech, M., P. Brown, J. Jones, A. R. Webster, and R. Worsfold. "Estimated Leonid Meteoroid Fluences Under Storm Conditions." In *Proc. of the 2nd European Conference on Space Debris*. Darmstadt, Germany, March 17–19, 1997 (ESA Sp-393, May 1997).

Ben-Shmuel, S. "Hypervelocity Shallow Angle Impact Simulation." In *Proc. of Second Leonid Storm and Satellite Threat Conference*. Manhattan Beach, CA: AIAA/Aerospace Corporation, May 1999.

Brown, P., K. J. Ellis, W. K. Hocking, A. R. Webster, and R. L. Hawkes. "Recent Observations and Modeling of the Leonid Meteoroid Stream." In *Proc. of First Leonid Meteoroid Storm and Satellite Threat Conference*. Manhattan Beach, California: AIAA/Aerospace Corporation, April 1998.

Brown, P., and J. Jones. "Evolution of the Leonid Meteor Stream." In *Meteoroids and Their Parent Bodies: Proc. of the International Astronomical Symposium*, Smolenice, Slovakia, July 6–12, 1992. Edited by J. Stohl and I. P. Williams, 57. Bratislava, Slovakia: Astronomical Institute of the Slovak Academy of Sciences, 1993.

Brown, P., P. J. Jones, K. J. Ellis, R. L. Hawkes, A. LeBlanc, A. R. Webster, M. Campbell, R. Correll, T. Montague, and S. P. Worden. "Observations of the 1998 Leonid Shower and Model Predictions of Shower Activity for 1999–2001." In *Proc. of Second Leonid Storm and Satellite Threat Conference*. Manhattan Beach, CA: AIAA/Aerospace Corporation, May 1999.

Brown, P., and J. Rendtel. "The Perseid Meteor Stream: Characterization of Recent Activity from Visual Observations." *Icarus* 124 (1996): 414.

Burns, J. A., P. L. Lamy, and S. Soter. "Radiation Forces of Small Particles in the Solar System." *Icarus* 40 (1979): 1–48.

Campbell, M. M., and D. F. Medina. "Kinetic Damage to External Satellite Materials Due to Impact of Leonid Micrometeoroids in the 10 Microgram Mass Range." In *Proc. of First Leonid Meteoroid Storm and Satellite Threat Conference*. Manhattan Beach, California: AIAA/Aerospace Corporation, April 1998.

Campbell, M. M., D. F. Medina, and L. A. Wright. "Solar Array Damage Predictions for Normal and Acute Angle Impacts by Sub-microgram Leonid Meteoroids." In *Proc. of Second Leonid Storm and Satellite Threat Conference.* Manhattan Beach, CA: AIAA/Aerospace Corporation, May 1999.

Carusi, A., L. Kresak, E. Perozzi, and G. B. Valsecchi. *Long-Term Evolution of Short-Period Comets*. Bristol, England, and Accord, Massachusetts: Adam Hilger, Ltd., 1985.

Caswell, D. "OLYMPUS and the 1993 Perseids, Lessons for the Leonids." In *Proc. of First Leonid Meteoroid Storm and Satellite Threat Conference.* Manhattan Beach, California: AIAA/Aerospace Corporation, April 1998.

Chobotov, V. A. *Orbital Mechanics*. Washington, D.C.: AIAA Education Series, 1991.

Christiansen, E. L. "Design and Performance Equations for Advanced Meteoroid and Debris Shields." *Int. J. Impact Engr.* 14 (1993): 145–156.

Cook, A. F. "A Working List of Meteor Streams." In *Evolutionary and Physical Properties of Meteoroids*. NASA SP-319, 1973.

Cooke, W. J., and B. J. Anderson. "Estimates of Leonid Storm Probabilities and Fluxes from Historical Information." In *Proc. of First Leonid Meteoroid Storm and Satellite Threat Conference.* Manhattan Beach, California: AIAA/Aerospace Corporation, April 1998.

Cooke, W. J., and R. M. Suggs. "The 1998 Leonid Apparition: Satellite Fluences and Projected Values for 1999." In *Proc. of Second Leonid Storm and Satellite Threat Conference.* Manhattan Beach, CA: AIAA/Aerospace Corporation, May 1999.

Danby, J. M. A. *Fundamentals of Celestial Mechanics*. Richmond, Virginia: Willman-Bell, Inc., 1989.

Dohnanyi, J. S. "Particle Dynamics." In *Cosmic Dust*. Edited by J. A. M. McDonnell. New York and Chichester, UK: Wiley, 1978.

Froeschle, C., and H. Scholl. "The Orbital Evolution of Meteor Streams at the 2-1 Resonance with Jupiter." In *Dynamics of Comets: Their Origin and Evolution*. Edited by A. Carusi and G. B. Valsecchi. Dordrecht, The Netherlands: D. Reidel Publishing Co., 1984.

Garrett, H. B., and A. C. Whittlesey. "Spacecraft Charging: an Update." In *Proc. of Second Leonid Storm and Satellite Threat Conference.* Manhattan Beach, CA: AIAA/Aerospace Corporation, May 1999.

Grün, E., H. Fechtig, H. A. Zook, and R. H. Giese. "Collisional Balance of the Meteoritic Complex." *Icarus* 62 (1985): 244–272.

Guth, V. *Bulletin of the Astronomical Institutes of Czechoslovakia* 1 (1) (1947): 171.

Hajduk, A. "Evolution of Cometary Debris—Physical Aspects." In *Comets in the post-Halley Era*, Vol. 1. Dordrecht, The Netherlands: Kluwer Academic Publishers, 1991.

Horanyi, M., and D. A. Mendis. "The Electrodynamics of Charged Dust in the Cometary Environment." In *Comets in the post-Halley Era*, Vol. 2. Dordrecht, The Netherlands: Kluwer Academic Publishers, 1991.

Hughes, D. W. "Meteors." In *Cosmic Dust*. Edited by J. A. M. McDonnell. New York and Chichester, UK: Wiley, 1978.

Ip, W.-H., H. Kmura, and C. P. Liu. "Interaction of the Cometary Dust with the Solar Wind and Cometary Plasma." In *Properties and Interactions of Interplanetary Dust*. Edited by R. H. Giese and P. Lamy. Dordrecht, The Netherlands: D. Reidel Publishing Co., 1985.

Jackson, A. A., and H. A. Zook. "Orbital Evolution of Dust Particles from Comets and Asteroids." *Icarus* 97 (1992): 70–84.

Jenniskens, P. "Meteor stream activity: I. Annual streams." *Astron. & Astrophys.* 287 (1994): 990–1013.

———. "Meteor stream activity: II. Meteor outbursts." *Astron. & Astrophys.* 295 (1995): 206–235.

———. "Meteor stream activity: III. Measurement of the First in a New Series of Leonid Outburst." *Meteoritics & Planetary Science* 31 (1996) 177–184.

———. "Meteor Stream activity: VI. Meteor Outbursts and the Reflex Motion of the Sun." *Astron. & Astrophys*. 317 (1997): 953–961.

———. "Meteor Stream activity: V. The Quadrantids, a Very Young Stream." *Astron. & Astrophys*. 327 (1997): 1242–1252.

Karrenberg, H. K. (private communication).

Kessler, D. *Space Station Program Natural Environment Definition for Design.* NASA SSP 30425, Revision A, 8:1, 1991.

Koschack, R., and J. Rendtel. "Determination of Spatial Density and Mass Index from Visual Meteor Observations (1)." *WGN, Journal of the IMO* 18 (1990a): 44–58.

———l. "Determination of Spatial Density and Mass Index from Visual Meteor Observations (2)." *WGN, Journal of the IMO* 18 (1990b): 119–140.

Kresak, L. "Evidence for Physical Aging of Periodic Comets." In *Comets in the post-Halley Era*, Vol. 1:607–628. Dordrecht, The Netherlands: Kluwer Academic Publishers, 1991.

Kronk, G., *Index of Comet Shower.* (web site) September 1998. Available: <www.maa.mhn.de/Comet/Shower> (current May 4, 1999).

Lamois-Cavelier, A. "KEOPS: Krystal European in Orbit Positioning System." Paper presented at *IAF Workshop on Satellite Constellations*, Toulouse, France, November 1997.

Lynch, D. K. "The Leonid Meteoroid Shower: Lessons Learned from 1997 and Plans for 1998." (unpublished).

———. "The Leonid Meteor Storm Threat." Paper presented at The Aerospace Corporation, El Segundo, California, September 1997.

Lynch, D. K., G. E. Peterson, and K. W. Jacobs. *Understanding the Leonid Meteor Storms*. (web site) September 1998. Available: <www.aero.org/leonid> (current at May 4, 1999).

Marsden, B. G. "Nongravitational Effects on Comets: the Current Status." In *The Motion, Evolution of Orbits, and Origin of Comets*. Edited by G. A. Chebotarev, E. J. Kazimirchak-Polonskaya, and B. G. Marsden. Dordrecht, The Netherlands: D. Reidel Publishing Company, 1970.

Marsden, B. G., and G. V. Williams. *Catalog of Cometary Orbits 1993*. 8th ed. Cambridge, Massachusetts, Smithsonian Astrophysical Observatory: 1993.

Matossian, M. G. "A Teledesic Space Infrastructure Overview." Paper presented at *IAF Workshop on Satellite Constellations*, Toulouse, France, November 1997.

McDonnell, J. A. M., P. L. Lamy, and G. S. Pankiewicz. "Physical Properties of Cometary Dust." In *Comets in the Post-Halley Era*, Vol. 2:1043–1073. Dordrecht, The Netherlands: Kluwer Academic Publishers, 1991.

McDonnell, J. A. M., N. McBride, and D. J. Gardner. "The Leonid Meteoroid Stream: Spacecraft Interactions and Effects." In *Proc. of the 2nd European Conference on Space Debris*. Darmstadt, Germany, 17–19 March 1997 (ESA Sp-393, May 1997).

McIntosh, B. A. "Origin and Evolution of Recent Leonid Meteor Showers; Perturbed Motion of Meteor Stream." In *Evolutionary and Physical Properties of Meteoroids*. NASA SP-319, 1973.

———. "Debris from Comets—the Evolution of Meteor Streams." In *Comets in the Post-Halley Era*, Vol. 1. Dordrecht, The Netherlands: Kluwer Academic Publishers, 1991.

Muraoka Laboratory Waseda University. (web site) November 1997. Available: <www.info.waseda.ac.jp/muraoka/> (current at May 5, 1999).

NASA Jet Propulsion Laboratory Solar System Dynamics Ephemeris Generator. (web site) October 1997. Available: <www.ssd.jpl.nasa.gov/cgi-bin/eph> (current at May 5, 1999).

Peterson, G. E. "Delta V Requirements and Optimal Node Locations for Earth Shadowing of the Leonid Meteors." In *Proc. of First Leonid Meteoroid Storm and Satellite Threat Conference*. Manhattan Beach, California: AIAA/Aerospace Corporation, April 1998.

———. "Dynamical Mitigation Strategies for Meteoroid Threat Minimization." In *Proc. of Second Leonid Storm and Satellite Threat Conference*. Manhattan Beach, CA: AIAA/Aerospace Corporation, May 1999.

Pizzicaroli, J. C. "Launching and Building the Iridium Constellation." Paper presented at *IAF Workshop on Satellite Constellations*, Toulouse, France, November 1997.

Porubcan, V., and J. Stohl. In *Asteroids, Comets, and Meteors*. Edited by A. W. Harris and E. Bowell. Houston: Lunar and Planetary Inst., 1991.

Ratcliffe, P. R., M. J. Cole, J. A. M. McDonnell, H. A. Shaw, and F. Allahdadi. "Experimental Determination of Energy Partitioning in Microparticle Impacts at Velocities from 1 to 100 km/s." *International Astronautical Congress, 46th*. Oslo, Norway: October 2–6, 1995.

Robinson, H. J. *Astronomy Data Book*. New York: Wiley & Sons, 1972.

Roy, A. E. *Orbital Motion*. New York: Adam Hilger Press, 1978.

Shahid-Saless, B., and D. K. Yeomans. "Relativistic Effects on the Motion of Asteroids and Comets." *Astron. J.* 107 (1994): 1885–1889.

Swift, H.F., "Hypervelocity Impact Mechanics," in *Impact Dynamics*. Edited by J.A. Zukas, T. Nicholas, H.F. Swift, L.B. Greszczuk, and D.R. Curran. Wiley & Sons, New York, 1982.

Tanner, W.G., C.R. Maag, W.M. Alexander, and P. Sappenfield, "Determination of Parameters for Hypervelocity Dust Grains Encountered in Near-Earth Space," *Proc. 24th Lunar and Planetary Science Conference*, Houston, Texas: Lunar and Planetary Institute, 1993.

Taylor, A. D., and N. McBride. "A Radiant-Resolved Meteoroid Model." In *Proc. of the 2nd European Conference on Space Debris*. Darmstadt, Germany, March 17–19, 1997.

Thomas, G. B., and R. L. Finney. *Calculus and Analytic Geometry*. Reading, Massachusetts: Addison-Wesley Publishing, 1980.

Vallado, D. A. *Fundamentals of Astrodynamics and Applications*. New York: McGraw-Hill, 1997.

Wasbauer, J. J., M. Blanc, F. Alby, and P. Cheoux-Damas. "Modeling Interplanetary Dust Distribution." In *Proc. of the 2nd European Conference on Space Debris. Darmstadt*, Germany, March 17–19, 1997.

Watts, A.J., D.R. Atkinson, S.R. Rieco, J.B. Lindvold, and S.L. Lapin, and C.R. Coombs, "A Fresh Look at Crater Scaling Laws for Normal and Oblique Hypervelocity Impacts," Proc. *24th Lunar and Planetary Science Conference*. Houston, Texas: Lunar and Planetary Institute, 1993.

Wertz, J. R., and W. J. Larson. *Space Mission Analysis and Design*. Dordrecht, The Netherlands: Kluwer Academic Publishers, 1991.

Weiss, A. A. "Radio-Echo Observations of Southern Hemisphere Meteor Shower Activity from 1956 December to 1958 August." *Monthly Notices of the Royal Astronomical Society* 120 (1960): 387–403.

Whipple, F. L. "Comets—Nature, Evolution and Decay." In *Highlights of Astronomy*, Vol. 6. Dordrecht, The Netherlands: D. Reidel Publishing Co., 1983.

———. "A Comet Model. I. The Acceleration of Comet Encke." *Astrophys. J.* 111 (1950): 375–394.

———. "A Comet Model. II. Physical Relation for Comets and Meteors." *Astrophys. J.* 113 (1951): 464–474.

Williams, I. P., and Z. Wu. "On the Quadrantid Meteoroid Stream Complex." *Monthly Notices of the Royal Astronomical Society* 259 (1992): 617.

———. "The Quadrantid Meteoroid Stream and Comet 1491 I." *Monthly Notices of the Royal Astronomical Society* 264 (1993): 659–664.

Yeomans, D. K. "Comet 55P/Tempel-Tuttle and the Leonid Meteors." In *Proc. of First Leonid Meteoroid Storm and Satellite Threat Conference*, Los Angeles, California, April 1998.

———. "Comet Tempel-Tuttle and the Leonid meteors." *Icarus* 47 (1981): 492–499.

Yeomans, D. K., and P. W. Chodas. "An Asymmetric Outgassing Model for Cometary Nongravitational Accelerations." *Astron. J.* 98 (1989): 1083–1093.

Yeomans, D. K., K. K. Yau, and P. R. Weissman. "The Impending Appearance of Comet Tempel-Tuttle and the Leonid Meteors." *Icarus* 124 (1996): 407–413.

# Index